Differential Equations on Fractals

Differential Equations on Fractals

A Tutorial

Robert S. Strichartz

PRINCETON UNIVERSITY PRESS

PRINCETON AND OXFORD

Library of Congress Cataloging-in-Publication Data

Strichartz, Robert S.

Differential equations on fractals: a tutorial / Robert S. Strichartz.

p. cm.

Includes bibliographical references and index.

ISBN-13:   978-0-691-12542-8 (cloth: alk. paper)

ISBN-10:   0-691-12542-2 (cloth: alk. paper)

ISBN-13:   978-0-691-12731-6 (pbk.: alk. paper)

ISBN-10:   0-691-12731-X (pbk.: alk. paper)

1. Fractals. 2. Differential equations. I. Title.

QA614.86.S77 2006

514′.742–dc22   2005057724

British Library Cataloging-in-Publication Data is available

This book has been composed in Times

Printed on acid-free paper. ∞

pup.princeton.edu

Printed in the United States of America

10 9 8 7 6 5 4 3 2 1

Contents

Introduction

There are more things in heaven and earth, Horatio,
than are dreamt of in your philosophy.

—*Hamlet* Act I scene V

It's a rough world out there. So why should all our mathematical models assume that the underlying space is smooth? In an attempt to grapple with this question, mathematicians are beginning to develop what might be called "rough analysis." For example, there is a large literature on PDEs on domains with nonsmooth boundary. But most of the results require assuming that the boundary is Lipschitz; this is just barely nonsmooth, since a Lipschitz boundary has a tangent plane almost everywhere. Another area of research is devoted to "metric-measure spaces." In a space equipped only with a metric and a measure, related by a doubling condition, it is possible to develop many results from classical harmonic analysis: Calderon–Zygmund theory, Hardy spaces, even Sobolev spaces of limited order of smoothness. But there are no differential equations in such a primitive context. There is also the beginning of a theory of "metric-measure-energy spaces," where by assuming an additional structure, an energy, or "local Dirichlet form," it becomes possible to define differential equations as well as stochastic processes. It is not the goal of these notes to develop such a grandiose theory. On the contrary, the aim is to look closely at a few basic examples. Our model is Apollonius, not Bourbaki.

The area of "fractal analysis" deals with analytic questions in which the underlying space is fractal. Here we want to distinguish two distinct approaches, which we will call "extrinsic" and "intrinsic." In the extrinsic approach the fractal set is embedded in a Euclidean space (or a manifold) and the analytic theory is that of the ambient Euclidean space. We mention just two examples: The book [Jonsson & Wallin 1984] develops a theory of function spaces on fractal subsets of $\mathbb{R}^n$ by restricting function spaces on $\mathbb{R}^n$ to the set, using ideas of the Whitney extension theorem; the existence of orthonormal bases of complex exponents with frequencies in a "spectrum," for certain Cantor measures in the line, was first proved in [Jorgensen & Pederson 1998]. In these notes we take the intrinsic approach, where the analytic structure arises from the fractal itself, and the embedding in Euclidean space is just a mental crutch for visualizing the domain.

The fractals we deal with lie at a polar extreme to smooth manifolds. While they may be regarded as rough objects, they also possess a very detailed structure called "self-similarity." It cannot be expected that objects in the real world, be they clouds, snowflakes, lightning bolts, blood vessels, or whatever, will possess such a rigid structure. Indeed, no physical object can be truly self-similar since once

Figure 1

we get to the molecular level the structure must change. So the construction of an analytic theory on certain self-similar fractals can only be regarded as a first step in the direction of an intrinsic fractal analysis. We can hope that it is a worthy first step, in the way that Apollonius's study of conics was a worthy first step in the understanding of general curves. At any rate, it is what we can do with the tools we have available now.

In the first three chapters of these notes we will deal with just two examples, the Sierpinski gasket (SG) (Figure 1) and the unit interval (I). The unit interval? Surely that is not a fractal set! But it is self-similar, being generated by the two mappings $F_0 x = \frac{1}{2} x$ and $F_1 x = \frac{1}{2} x + \frac{1}{2}$ that make the two half-intervals $[0, \frac{1}{2}]$ and $[\frac{1}{2}, 1]$ similar to the whole interval. SG (sometimes called the "Sierpinski triangle") is generated by three mappings in the plane, each a similarity with ratio $\frac{1}{2}$ and fixed points the vertices of a triangle. Thus SG is a union of three smaller copies of itself (self-similarity), and these copies intersect each other at a finite set of points (finite ramification). So there is a strong kinship between these two examples. We will exploit this kinship in the following way: We start with a known analytic structure on I, find a way to reformulate it in terms of the self-similar structure, and then try to generalize the reformulated structure to SG. In the process we will actually gain new insight into analysis on I, although we will not actually prove new results.

There are two approaches to defining a Laplacian on SG. The first uses probability theory to define a stochastic process, analogous to Brownian motion, and the Laplacian emerges fully formed, but not in a truly constructive manner. This was historically the first to succeed, in works such as [Goldstein 1987], [Kusuoka 1987], [Barlow & Perkins 1988], and [Lindstrøm 1990]. The notes by [Barlow 1998] present the subject from this point of view. The advantages of this approach are that it is well suited to yield heat kernel estimates, an important tool that so far is not obtainable by the second method, and it is extendible to a class of Sierpinski carpets. The disadvantages are that it demands of the reader a strong background in probability theory, and it doesn't provide a method to actually do computations.

The second approach, based on calculus, is due to Jun Kigami. His monograph [Kigami 2001] gives a detailed and formal presentation of the theory in its greatest generality. These notes may be regarded as a soft introduction. (Expository accounts are given in [Kigami 1994b] and [Strichartz 1999b].) In calculus, operations on functions are defined as limits of discrete analogs. In addition to a general theory, there are numerous examples where the limits may be computed explicitly.

On SG and related fractals, the basic operations (energy, Laplacian, normal derivative) will be defined as limits of discrete operations on a sequence of graphs whose
vertices approximate the fractal. Moreover, we will see how to compute these limits
for particular functions.

. We should point out a technical matter that limits the extent of the possible scope
of this approach. It is the fact that *points have positive capacity*. This is a cluster
of equivalent conditions, but basically it allows us to work entirely with continuous
functions. So, in particular, the restrictions of functions to discrete sets are well
defined, and a function is determined by its restriction to a dense subset. This is not
the situation when it comes to the square, for example. There are functions of finite
energy,

$$\int_0^1 \int_0^1 |\nabla u(x, y)|^2 dx dy < \infty,$$

that are discontinuous. So we cannot obtain the energy as a limit of discrete energies. Of course, this is not a fatal flaw, since the square is the product of intervals,
so we can use a product construction to transfer analysis from the interval to the
square.

As hinted above, the central concept in this theory is energy. Although energy is
supposed to be the analog of the integral of the square of the gradient, in the fractal
case it will be constructed in one step, because essentially there is no gradient or
integral in our energy. (This is not to say that one cannnot try to push through
such notions after the fact. See [Kusuoka 1989].) This construction is at the center
of Chapter 1. First we discuss the self-similar structure of our fractal and self-
similar measures on the fractal. At the end of the chapter we discuss the effective
resistance metric, defined in terms of the energy. By the end of the chapter we
have assembled the three essential ingredients for our analysis: measure, energy,
and metric. We also discuss *harmonic functions*, defined to be energy minimizers
(subject to boundary conditions), the generalization to SG of linear functions on
I. The space of harmonic functions is three-dimensional, and there is a simple
harmonic extension algorithm, the "$\frac{1}{5} - \frac{2}{5}$ rule," that enables us to compute the
values of a harmonic function on a finer level from values on a coarser level.

In Chapter 2 we introduce the main protagonist: the Laplacian. The two main
ingredients in the construction are the energy and a measure. In particular, the same
energy can give rise to different Laplacians, one for each choice of measure. In our
examples there is a standard measure, self-similar and symmetric with respect to
the symmetries of the fractal, and this gives rise to a standard Laplacian. However,
in the most general context of this theory it is more difficult to justify the choice
of a standard measure. The idea of a whole family of Laplacians is at first a bit
disturbing. On Riemannian manifolds the Riemannian metric gives rise to all three
structures: metric, energy, and measure, so there is a unique Laplacian (Laplace–
Beltrami operator). On the other hand, there is an extensive theory of second-order
elliptic operators that is even more general. In our context there is a way of encom-
passing all the Laplacians at once, by having the Laplacian of a function be a signed
measure. This has the disadvantage that we cannot apply the Laplacian a second

time. Since this idea uses more technical results from measure theory, we relegate it to the exercises.

We give two equivalent definitions of the Laplacian: a weak formulation and a pointwise formula. The pointwise formula for the standard Laplacian resembles the second difference quotient formula for the second derivative, with the constant 4 replaced by 5. In our presentation, the constant 5 emerges as the product of two others, one associated to energy and one associated to the measure. We deliberately put the weak formulation first, since we believe it is more natural, and it can be used in computations. A slight extension of the weak formulation leads to the Gauss–Green formula, a great workhorse in the further development of the theory. Before we can state the Gauss–Green formula, however, we need to define and study normal derivatives at boundary points. In fact, the definition of normal derivatives depends only on the energy, not the Laplacian. However, the main result we use is that a function in the domain of the Laplacian (using any measure) has normal derivatives. The normal derivatives can be localized to some nonboundary points, and they provide a criterion for gluing together functions defined in pieces to obtain functions in the domain of the Laplacian.

The definitions of the Laplacian do not come with any guarantee that there are any nonconstant functions in the domain of the Laplacian. This leaves us with the disquieting possibility that we are describing a vacuous theory (like Hölder continuous functions on the line with exponent greater than 1). We establish early on that harmonic functions (those minimizing energy with given boundary values) have Laplacian equal to zero (for any Laplacian), but on SG the harmonic functions form only a three-dimensional space. What finally saves the day is the Green's function, the kernel of the inverse of the Laplacian (actually $-\Delta$) with Dirichlet boundary conditions. We give an explicit construction of this function of two variables. The construction shows that the Green's function is continuous, one of the cluster of equivalent conditions that points have positive capacity. With this we know that the domain of the Laplacian is as large as it could possibly be, since we can solve $-\Delta u = f$ for every continuous function f. The construction itself is very interesting, making use of the self-similar structure and yielding a new method for obtaining the Green's function on the interval.

Chapter 2 concludes with a detailed description of the local behavior of "smooth" functions. We do this first for harmonic functions, where the results are very precise, and then for functions in the domain of the Laplacian, where details are a bit fuzzier. Here we discover another local derivative, called the tangential derivative. At a boundary point, the data consisting of the value of the function, its normal derivative, and its tangential derivative plays the role of a 1-jet in smooth analysis. We also discover a negative and perhaps disturbing fact: The domain of the Laplacian does not form an algebra under multiplication! Actually, the situation is much worse: If u is a nonconstant function in the domain of the Laplacian, then u^2 is not in the domain of the Laplacian (the same is true for other nonlinear functions of u, as well). This situation is not unprecedented in smooth analysis, as it is the case for the domain of an elliptic operator with nonsmooth coefficients.

The Laplacian on SG is not a differential operator in the usual sense of the term. It may appear that since SG contains many line segments, one could easily take ordinary directional derivatives by looking at the restriction of a function to those line segments. But this idea leads to a dead end. A generic point on SG does not lie on any of these line segments. Moreover, there is no effective way to combine all these directional derivatives into a coherent structure. The functions in the domain of our Laplacian do not have meaningful directional derivatives. Also, more general fractals where our method works do not necessarily contain any rectifiable curves. But our Laplacian is a local operator, and it is the limit of suitable difference quotients, so it seems reasonable to call it a "differential operator."

It does not follow that it is a "second-order" differential operator. Although this supposition has been tacitly used in the literature to motivate certain definitions, the evidence clearly points in a different direction. We do not have a differential calculus on SG that would give an inevitable concept of "order" for differential operators. But there are many classical results, such as the Sobolev embedding theorem, where the order of a differential operator makes an appearance. When we look at the analogous results on SG, we see a consistent pattern with the number $d + 1$ standing in for the order, where d is the dimension of SG in the effective resistance metric. Note that this is perfectly consistent with what happens on the interval, where $d = 1$ and $d + 1 = 2$.

In Chapter 3 we discuss the spectrum of the standard Laplacian on SG, using the method of spectral decimation. The idea here was first considered by physicists, but the mathematical results come from [Fukushima & Shima 1992]. This method allows us to describe all eigenfunctions, regardless of boundary conditions, as well as the exact Dirichlet and Neumann eigenvalues and eigenfunctions. The eigenfunctions are described by a variant of the harmonic extension algorithm, but with a new twist that each eigenfunction has a "generation of birth," a level at which the construction must begin fresh.

To motivate the construction on SG we begin by presenting the corresponding construction on the interval. The eigenfunctions there are the familiar exponentials, sines, and cosines. Not only are these eigenfunctions of the continuous Laplacian, but they are also eigenfunctions of the discrete Laplacians, although the eigenvalues are different. This is of course well known, but what is perhaps not so well known is that the entire structure of these functions can be built up from this simple observation. After we do the method on I, we carry out the generalization to SG. First we deal with the case of "generic" eigenvalues, where there is no obstruction to the extension algorithm. Then we deal with Dirichlet and Neumann eigenfunctions, and these correspond to nongeneric eigenvalues. Some of these eigenvalues have very high multiplicity, and the eigenspaces contain functions that are highly localized (have small support). These properties go hand in hand with another property: the existence of joint Dirichlet–Neumann eigenfunctions. Although these properties are quite apparent from the spectral decimation method, we show that they may be traced to the dihedral symmetry of SG.

The actual spectrum (Dirichlet or Neumann) of the standard Laplacian is very lumpy. In addition to the large multiplicities (relative to the total number of eigenvalues up to that point), there are infinitely many large gaps (relative to the eigenvalue). We may define an eigenvalue counting function and look for a Weyl-type law with an appropriate power giving its asymptotics, but this is not precisely the case: The power must be multiplied by a periodic function (in the logarithm of the eigenvalue), and moreover this function has jump discontinuities. One clue to this behavior is the renormalization polynomial that describes the eigenvalues of the discrete Laplacians (going from the finer to the coarser level). For the interval the polynomial is $x(4-x)$, whose Julia set is the interval $[0, 4]$. For SG the polynomial is $x(5-x)$, whose Julia set is a Cantor set in $\mathbb{R}$. In a certain sense, the limit at infinity of the spectrum is equal to the limit at zero of the Julia set. Also note that once again the difference between I and SG is the substitution of 5 for 4. For now this appears to be just a coincidence.

In order to describe a complete analog of Fourier series on SG we need to find an orthonormal basis of eigenfunctions. This creates considerable difficulty because of the eigenspaces of high multiplicity. It is easy enough to describe a basis for these eigenspaces, but there do not appear to be any natural orthonormal bases. What we are able to do is give a formula for the inner product for each eigenspace, in terms of a discrete inner product at the generation of birth. This formula is complicated, involving an infinite product of terms based on the discrete eigenvalues at all levels, but it is reasonably explicit.

In Chapter 4 we move on to a larger class of self-similar fractals for which it is possible to develop an analogous theory of energy and Laplacian. We define a class of postcritically finite (pcf) self-similar fractals (not quite in the greatest generality). These are "finitely ramified" in the sense that they are both connected and close to being disconnected—the removal of a finite number of strategically chosen points is all it takes to disconnect them. It is possible to approximate pcf fractals by a sequence of graphs, and so we can try to find a sequence of discrete energies that will converge to a self-similar energy on the fractal. There are more degrees of freedom here than in the case of SG, so the problem is much harder. We define a restriction map that takes us from energy on a finer approximation to energy on a coarser approximation (essentially just minimize the energy over all possible extensions of a function). When we combine this with the self-similar structure we obtain a renormalization map from a finite-dimensional space (the parameter space for all possible discrete energies) to itself. This map is nonlinear. What we need is a nondegenerate eigenvector, for then we can take the corresponding energy as our initial discrete energy, and the eigenvalue as a renormalization factor in going from one level of approximation to the next. The self-similar structure says that if we can solve this renormalization problem at one level, then the same solution works on all levels.

The subtle issue here is that we require that the eigenvector be nondegenerate, not merely nonzero. Essentially we will require that all entries be strictly positive. It is perhaps easier to understand what we don't want. Look at SG and one of the line segments joining two boundary points. If we take the usual energy on this interval, we could declare it to be a self-similar energy on SG. It will have all the usual properties of energy except one: There are nonconstant functions with zero energy. So it is a kind of degenerate energy. This energy is easily constructed by the method sketched above, but it starts from a discrete energy that is also degenerate.

If we didn't need to impose the nondegeneracy condition, we could easily prove the existence of solution to the renormalization problem using the Brouwer fixed point theorem. In general, it is not known whether or not solutions exist. In some examples it is known that solutions are not unique. Fortunately, there are many examples where solutions are known explicitly. There is a large class of fractals introduced in [Lindstrøm 1990], called *nested fractals*, possessing dihedral symmetry. For these fractals there is always a unique self-similar energy sharing these symmetries. Proofs of existence and uniqueness are rather technical and are not presented here.

Most of Chapter 4 is devoted to showing what can be done under the assumption that there is a solution to the renormalization problem. This includes most of the material in Chapters 1 and 2. Spectral decimation is not part of the story ([Shima 1996] gives a class of fractals where it holds, but it is a very restrictive class). We also discuss the "geography is destiny" principle, which is easier to illustrate using our wider class of examples. Although these self-similar fractals exhibit repetitive patterns at all scales, they are very far from being homogeneous. There are indeed lots of local isomorphisms, but they do not extend to global isomorphisms, and that matters. Specifically, if we take certain natural classes of functions, such as harmonic functions or functions in the domain of the Laplacian, and restrict these functions to small cells, the restricted functions might not be identical, even though the cells are locally isomorphic. Their location in the fractal (geography) influences what functions look like (destiny).

This is especially clear for harmonic functions, because they form a finite-dimensional space. On SG the space of harmonic functions is three-dimensional, and the restriction to any small cell always gives the same three-dimensional space. But much of this space is "improbable." Each cell (well, almost every one) has a certain two-dimensional subspace that it "favors," in the sense that if you choose a global harmonic function at random and restrict to that cell, you end up in the two-dimensional subspace with very high probability. This is a rather subtle phenomenon and is essentially a consequence of the theory of products of random matrices. But on other fractals the effect is more blatant: The space of restrictions of harmonic functions might have a smaller dimension than the space of global harmonic functions. This implies that there exist nonconstant harmonic functions that are locally constant.

We also discuss some examples of fractals that are not strictly self-similar but still fit into the pcf framework. These were introduced in [Hambly 1992, 1997] as examples of "random fractals," but in fact a good bit of the theory is valid for every example, not just almost every example. These examples show that it is the hierarchical structure of the fractal that is crucial, not necessarily the repetition of the same construction at all scales.

In Chapter 5 we touch lightly on a variety of topics. Some of them have been hinted at earlier or have appeared in the exercises. Some of them are of recent origin. The presentation is more sketchy than in the earlier chapters. The goal is to present the reader with a panoramic view of the field as it is being actively developed.

These notes are written in a deliberately informal style, with an emphasis on motivations. Sometimes we find it convenient to give formal statements of theorems and proofs, but we often omit details that are routine or leave them to the exercises.

We don't always take the shortest route to a goal, and we add interesting comments whenever possible.

These notes are conceived as a conversation with the reader. What is the reader's part in the conversation? The exercises. There are a large number of exercises, with varying degrees of difficulty. Those marked with an asterisk are either of greater difficulty or require some background material beyond the minimum needed to follow the text. It is strongly recommended that readers work some exercises in each section before going on to the next section, to test understanding and reinforce the ideas discussed.

Each chapter ends with a section of "notes and references," where we attempt to give correct attribution of the ideas presented and refer the reader to sources of further information. These comments are separated out from the rest of the text so as not to clutter the presentation.

These notes are directed at readers with different levels of preparation. This approach is not usually recommended, as it risks confusing the novices and at the same time boring the experts. Nevertheless, we feel that it is important to try to reach many audiences. We hope these notes will be easily understood by undergraduate students who have had a standard real analysis course, especially if they are willing to accept on trust certain technical facts. (Indeed I have been teaching this material to groups of students at this level for many years, and some of them have gone on to make contributions that are discussed in the present notes.) On the other hand, there are research mathematicians in related areas who might be intrigued by fractal analysis or tempted to work on problems in this area. We hope these notes will provide such readers with an informative introduction and that some of the ideas discussed here will resonate with ideas already known from related areas.

We hope that all readers will find something new to dream of in their philosophy.

Acknowledgments

I would like to thank Kasso Okoudjou, Roberto Peirone, Mark Pinsky, Luke Rogers, and Maria Roginskaya for pointing out errors in the preliminary version of the text, and Shawn Drenning and Alexander Teplyaev, who provided some of the figures.

I am grateful to the National Science Foundation, both for support of my research and for support of the many undergraduate students who have worked with me in the Research Experiences for Undergraduates (REU) Program at Cornell. It has been a pleasure to work with so many enthusiastic and talented students, and they have made essential contributons to the development of the field.

My deepest gratitude goes to Jun Kigami and Alexander Teplyaev, who have helped me to understand the material presented here. Through many conversations over the years they have been with me, pointing out some of my mistakes and confusions, encouraging my explorations, and supplying key ideas.

I am grateful to Arletta Havlik for skillfully preparing the TeX version of this book from my handwritten manuscript, as she has done for so many of my papers.

Ithaca, NY
August 2005

Differential Equations on Fractals

Chapter One

Measure, Energy, and Metric

1.1 GRAPH APPROXIMATIONS

In ordinary calculus we learn that continuous structures may be approximated by discrete structures. For example, the derivative is the limit of difference quotients, the integral is the limit of Riemann sums, and so on. At first, our naive intuition is that the discrete structures are simpler than the continuous ones, but we soon learn otherwise: The rules for derivatives are simpler than the corresponding rules for difference quotients (in fact, such rules as the product rule are rarely stated explicitly for difference quotients, although they do underlie the proofs of the corresponding derivative rules), and the fundamental theorem of the calculus allows very easy evaluation of some integrals. As we study calculus on fractals, we will also take the approach of using discrete approximations. At present there are no results that make the continuous structures simpler than the discrete ones, so we will have to devote careful attention to the discrete case. In the process we will learn some new things about ordinary calculus, since the unit interval is itself a self-similar set. Our plan is to develop the theory simultaneously for two examples: the unit interval I and the Sierpinski gasket SG.

The usual definition of the derivative involves arbitrary increments, and the Riemann sums in the definition of the integral allow arbitrary subdivisions of the interval. This is unnecessarily complicated. It suffices to deal with dyadic points $k/2^m$ ($0 \le k \le 2^m$, $0 \le m < \infty$). These points are dense in the interval, and as long as all the functions we deal with are continuous, it suffices to know the values at the dyadic points. To see how the dyadic points arise naturally we need to examine the self-similar structure of I. Consider the mappings $F_0 x = \frac{1}{2}x$ and $F_1 x = \frac{1}{2}x + \frac{1}{2}$ that send I to its left and right halves. Note that these are both contractive similarities (contraction ratio $\frac{1}{2}$, fixed points 0 and 1, respectively) and the images $F_0 I$ and $F_1 I$ intersect at the point $\frac{1}{2}$. The self-similar identity (the whole as union of similar parts)

$$(1.1.1) \qquad I = F_0 I \cup F_1 I$$

uniquely determines I, as long as we specify that I is a nonempty compact set (both the empty set and the whole line satisfy (1.1.1), as well as the set of rational numbers in I, dyadic numbers in I, etc.). We note that (1.1.1) is not the only self-similar identity for I. For example, we can get many more by iteration. Write $F_w = F_{w_1} \circ F_{w_2} \circ \cdots \circ F_{w_m}$ for $w = (w_1, \ldots, w_m)$, each $w_j = 0$ or 1 (we call w a *word* of length $m = |w|$). Then

$$(1.1.2) \qquad I = \bigcup_{|w|=m} F_w I$$

holds for any m. We will call this the *level m decomposition* and call $F_w I$ a *cell of level m*. Of course, (1.1.2) is just the decomposition of I into dyadic intervals $[k/2^m, (k+1)/2^m]$. We could also do irregular decompositions, such as

$$(1.1.3) \qquad\qquad I = F_0 I \cup F_{10} I \cup F_{11} I.$$

There are also totally unrelated self-similar identities, for example involving $\frac{1}{3}x$, $\frac{1}{3}x + \frac{1}{3}$ and $\frac{1}{3}x + \frac{2}{3}$. This shows that the interval is different from the other fractals we will be studying.

The dyadic points are just the boundary points of the cells of various levels. Let us introduce some notation. $V_0 = \{q_0, q_1\}$ for $q_0 = 0$ and $q_1 = 1$ is the set of boundary points of I. Then inductively

$$(1.1.4) \qquad\qquad V_m = \bigcup_i F_i V_{m-1},$$

or equivalently

$$(1.1.5) \qquad\qquad V_m = \bigcup_{|w|=m} \bigcup_i F_w q_i,$$

give the set of dyadic points $\{k/2^m\}$ for fixed m. Note that aside from the boundary points V_0, every point in V_m has two addresses, $x = F_w q_0 = F_{w'} q_1$, for the appropriate choices of w and w', so x is the left endpoint of one cell and the right endpoint of an adjacent cell. We will call such points *junction points*. We will regard the sets V_m as the vertices of a graph Γ_m, with edges written $x \underset{m}{\sim} y$ provided $x = k/2^m$ and $y = (k+1)/2^m$. Equivalently $x \underset{m}{\sim} y$ if there exists a cell of level m containing both x and y (as boundary points). Inductively, we build the graph Γ_m from the graph Γ_{m-1} by taking the two images $F_0 \Gamma_{m-1}$ and $F_1 \Gamma_{m-1}$ and identifying the common vertex $\frac{1}{2}$. See Figure 1.1.1. Note that the set of vertices is increasing,

$$(1.1.6) \qquad\qquad V_0 \subseteq V_1 \subseteq V_2 \subseteq \cdots.$$

However, the edge relations change: If x, y both belong to V_m and $x \underset{m}{\sim} y$, then x, y both belong to V_{m+1} but are *not* connected by an edge in Γ_{m+1}. Also note that every junction point in Γ_m has exactly two neighbors in V_m. Of course these graphs are very boring!

So now let's look at the case of SG. The self-similar structure of SG may be viewed as a natural generalization of the self-similar structure of I. This time we

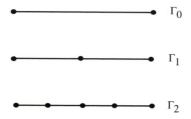

Figure 1.1.1

work in the plane and consider a set of three mappings $F_i : \mathbb{R}^2 \to \mathbb{R}^2$, $i = 0, 1, 2$, defined by

$$(1.1.7) \qquad\qquad F_i x = \frac{1}{2}(x - q_i) + q_i,$$

where $\{q_i\}$ are the vertices of a triangle (any nondegenerate triangle will do). Then SG satisfies the self-similar identity

$$(1.1.8) \qquad\qquad SG = \bigcup_{i=0}^{2} F_i(SG).$$

As in the case of I, the mapping F_i is a contractive similarity with contraction ratio $\frac{1}{2}$ and fixed point q_i. Also, SG is the unique nonempty compact set satisfying (1.1.8). The three cells on the right side of (1.1.8) intersect pairwise at single points. This means that while SG is connected, it is just barely so. If you remove just these three junction points, it becomes disconnected. You could think of SG as the ideal police state. To keep track of the whereabouts of all its citizens (at this level), the state need only post sentries at these three points. Similarly, if the state wants more detailed locations, it will post a finite number of sentries at more junction points. The terms "finitely ramified" and "postcritically finite" are used to describe this topological property. We will discuss the latter term in Chapter 4.

It is important to keep in mind that it is only the topological structure of SG that is of interest here, not the geometric structure inherited from its embedding in the plane. That is why we don't care which triangle we start with. But there are many other embeddings of SG in the plane, such as the famous Apollonian packing. We don't want to prejudice ourselves by looking at SG with "Euclidean eyes." In particular, although SG contains many straight line segments, we don't use any ordinary calculus concepts obtained by restricting functions on SG to these line segments. Eventually we will introduce a natural metric on SG that is not equivalent to the Euclidean metric (in any embedding in any dimensional Euclidean space) and that contains no rectifiable curves. Also, although our Euclidean eyes tend to see the triangle containing SG as a sort of boundary (since SG has no interior, the topological notion of boundary is not relevant), we will *define* the boundary to be the set $\{q_0, q_1, q_2\}$ of vertices of the triangle.

In order to be able to discuss I and SG simultaneously, we will use the symbol K to denote either one (and later other self-similar sets). The self-similar identities (1.1.1) and (1.1.8) may be combined as

$$(1.1.9) \qquad\qquad K = \bigcup_i F_i K$$

(taking advantage of the ambiguity concerning the number of terms in the union). By iteration we have

$$(1.1.10) \qquad\qquad K = \bigcup_{|w|=m} F_w K,$$

where F_w is defined as before, but the letters w_j in the word w may take on the values $\{0, 1, 2\}$ in the case of SG. This will be our decomposition of K into cells of

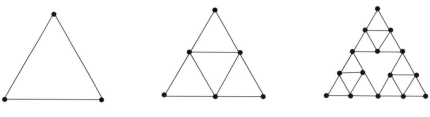

Figure 1.1.2

level m. Note that in the case of SG as well as I, distinct cells of level m are either disjoint or intersect at a single point; these will be our junction points. In the case of SG, unlike the interval, the junction points are topologically distinguishable from general points. In other words, while there are analogs of the decomposition (1.1.3), there are essentially no other decompositions. For SG we take $V_0 = \{q_0, q_1, q_2\}$. Then (1.1.4) or (1.1.5) defines V_m in both cases. Note that every point in $V_m \setminus V_0$ is a junction point with two addresses, $x = F_w q_i = F_{w'} q_{i'}$ for the appropriate choices (always $i \neq i'$). In particular,

$$(1.1.11) \qquad \{F_0 q_1 = F_1 q_0, \, F_1 q_2 = F_2 q_1, \, F_2 q_0 = F_0 q_2\}$$

are the three junction points in $V_1 \setminus V_0$ in the SG case. We construct a graph Γ_m with vertices V_m by defining the edge relation $x \underset{m}{\sim} y$ if there is a cell of level m containing both x and y ($\exists\, w$ with $|w| = m, i, j$ such that $x = F_w q_i$ and $y = F_w q_j$). In the case SG, Γ_m is obtained by taking the three copies $F_i \Gamma_{m-1}$ of Γ_{m-1} and identifying the points (1.1.11). Figure 1.1.2 shows the first three graphs.

Note that on SG every vertex in V_m, except for the three boundary points, has exactly four neighbors in Γ_m. There are times when the existence of the boundary is technically annoying, but we can easily get rid of it by passing to the double cover $\widetilde{K}$. That is, we take two copies of K and glue them together at the common boundary points. If we do this for I we obtain the circle, a one-dimensional manifold without boundary. The glued boundary points in $\widetilde{SG}$ have neighborhoods that are homeomorphic to neighborhoods of any junction point in SG. In this way, $\widetilde{SG}$ is an example of a "fractafold" without boundary modeled on SG. In the graphs $\widetilde{\Gamma}_m$ (two copies of Γ_m glued together at the corresponding boundary points), every vertex has exactly four neighbors. In other words, the graph is 4-regular.

EXERCISES

1.1.1. Show that $\#V_m = \frac{1}{2}(3^{m+1} + 3)$.

1.1.2. Let $x = F_w q_i$ with $|w| = m$ in $V_m \setminus V_0$. Give an algorithm for finding the other address $F_{w'} q_j$ for x in V_m. (Hint: If $w_m \neq i$, then $w'_k = w_k$ for $k < m$, $w'_m = i$, and $j = w_m$. If $w_m = i$, then $x \in V_{m-1}$, so reason by induction.)

1.1.3. Explicitly identify the four neighbors of x in Γ_m for $x \in V_m \setminus V_0$ and the two neighbors of q_i.

1.1.4. Show that the dihedral symmetry group D_3 of the equilateral triangle (three reflections, two rotations, and the identity) acts as a symmetry group of SG, and the action on Γ_m is given by permutations of the letters $\{0, 1, 2\}$.

1.1.5. Show that the set $V_1 \setminus V_0$ in SG is characterized topologically as the only set of three points whose removal disconnects SG into three components.

1.1.6. Show that SG is topologically rigid: Any homeomorphism must be one of the six symmetries in D_3.

1.1.7.* Show that $\widetilde{SG}$ is not topologically rigid in that there are infinitely many "accordian moves" across an identified boundary point.

1.1.8. Show that any two points in SG may be joined by a rectifiable curve (in fact, an infinite polygonal line).

1.1.9. (Nesting property) Show that if two cells (not necessarily of the same level) intersect in more than one point, then one contains the other.

1.1.10. If $x \in V_m$, then there exists a "chain" of points (not necessarily distinct) $x_0, x_1, \ldots, x_m$ such that $x_0 = q_0$, $x_k \in V_k$, and $x_m = x$, and $x_{k-1} \underset{k}{\sim} x_k$ for $1 \le k \le m$.

1.2 SELF-SIMILAR MEASURES

The notion of a general measure is a far-reaching generalization of notions such as length, area, volume, and probability. The theory is quite technical, and we will not attempt to describe it here. If you are familiar with measure theory, then you will be able to understand what we do here in that broader context (the key tool is the extension theorem of Carathéodory). If you are not familiar with measure theory you can still relax, because everything we are going to do is quite simple. This is thanks to the self-similar structure of K and also to the fact that we only need to integrate continuous functions, so we may imitate the integrals of Cauchy and Riemann rather than the integral of Lebesgue.

We want to consider what will be called here a *regular probability measure* μ on K. Roughly speaking, μ assigns weights $\mu(C)$ to all cells C of K in an additive fashion. Precisely, we require just the following four conditions:

(1.2.1) (positivity) $\mu(C) > 0$;

(1.2.2) (additivity) if $C = \bigcup_{j=1}^{N} C_j,$

where the cells $\{C_j\}$ intersect only at boundary points, then

$$\mu(C) = \sum_{j=1}^{N} \mu(C_j);$$

(1.2.3) (continuity) $\mu(C) \to 0$

as the size of C goes to 0;

(1.2.4) (probability) $\mu(SG) = 1$.

Condition (1.2.3) says that points have zero measure, and this enables us to ignore point intersections in condition (1.2.2). We may then extend the domain of μ to include finite unions of cells: If

(1.2.5) $$A = \bigcup_{j=1}^{N} C_j$$

is a finite union of cells, disjoint except for point intersections, define

(1.2.6) $$\mu(A) = \sum_{j=1}^{N} \mu(C_j).$$

Condition (1.2.2) guarantees that this is unambiguous: A different decomposition into cells would yield the same measure. To see this, first observe that there is a unique canonical decomposition of A into a union of cells of maximal size (C is not contained in a larger cell in A). Indeed, because of the nesting property, the maximal cells in A are disjoint (except for point intersections) and their union is A. Then any other representation of A is obtained by subdividing the maximal cells in some manner, and (1.2.2) says that the measure is conserved in the process. The additivity condition (1.2.2) continues to hold for sets that are finite unions of cells.

Similar reasoning shows that in place of (1.2.2) it suffices to verify for every cell $F_w K$,

(1.2.7) $$\mu(F_w K) = \sum_i \mu(F_w F_i K),$$

the additivity for the decomposistion of $F_w K$ into cells of the next level. The construction of μ can then be imagined as follows: We assign weight 1 to SG, the cell of level 0. Inductively, having assigned weights to all cells of level m, we decide how to split the weight of each such cell when we subdivide it into cells at level $m + 1$. The only restrictions are that (1.2.7) and (1.2.1) hold for the splitting, and (1.2.3) holds in the limit. Clearly there is a huge selection of measures!

The simplest choice is to do all splittings evenly. In the case of I, each cell of level m has measure 2^{-m}, its usual length. In fact, if A is any interval with dyadic endpoints, then $\mu(A)$ is the length of A. In the case of SG, each cell of level m will have measure 3^{-m}. We refer to this as the *standard measure*. It happens to coincide, up to a constant, with Hausdorff measure on SG in dimension $\log 3 / \log 2$ (the exact value of the constant is an unsolved problem). Of course, the definition of Hausdorff measure is quite complicated, whereas our definition is quite simple.

The standard measure is a special case of a *self-similar* measure. To determine a self-similar measure we choose a set of probability weights $\{\mu_i\}$ on the index set $\{0, 1\}$ or $\{0, 1, 2, \}$,

(1.2.8) $$\sum_i \mu_i = 1 \quad \text{with } \mu_i > 0,$$

and then set

(1.2.9) $$\mu(F_w K) = \prod_{j=1}^{m} \mu_{w_j} \quad \text{for } |w| = m.$$

For simplicity we will write μ_w for the right side of (1.2.9). Another way of saying this is that we use the weights $\{\mu_i\}$ to accomplish the splitting (1.2.7). The standard measure is obtained by choosing all the μ_i equal, so $\mu_i = \frac{1}{2}$ for I and $\mu_i = \frac{1}{3}$ for SG.

For a self-similar measure, each mapping F_i contracts measures of sets by a factor μ_i,

$$(1.2.10) \qquad \mu(F_i A) = \mu_i \mu(A),$$

since this is clearly true for cells by (1.2.9). Another way of expressing this is to take a set A and split it as $\bigcup_i A \cap F_i K$, and then by additivity,

$$(1.2.11) \qquad \mu(A) = \sum_i \mu(A \cap F_i K).$$

But $F_i^{-1} A = F_i^{-1}(A \cap F_i K)$ and $A \cap F_i K = F_i F_i^{-1}(A \cap F_i K)$, so $\mu(A \cap F_i K) = \mu_i \mu(F_i^{-1}(A \cap F_i K))$ by (1.2.10). Together this shows $\mu(A \cap F_i K) = \mu_i \mu(F_i^{-1} A)$. Substituting into (1.2.11) yields the self-similar identity

$$(1.2.12) \qquad \mu(A) = \sum_i \mu_i \mu(F_i^{-1} A).$$

It is easy to see that (1.2.12) implies (1.2.10) (replace A by $F_i A$, and then only one term survives on the right side) and hence (1.2.9), so the self-similar identity (1.2.12) (and the probability condition (1.2.4)) determines the measure μ uniquely. (It is also possible to prove this using a form of the contractive mapping principle, but the argument is more technical.)

On I, the self-similar measures are often called *Bernoulli measures*. Using the binary expansion, we may identify $x \in I$ with an infinite string of 0's and 1's (there is some ambiguity when x is a binary rational, but the set of binary rationals has measure zero and can be ignored). If we choose 0 with probability μ_0 and 1 with probability μ_1, independently for each binary digit, we get exactly the self-similar measure. Similarly, on SG we can interpret a self-similar measure as giving a recipe for choosing a point x in SG "at random." We first decide which of the three cells $F_i K$ of level 1 the point belongs to by spinning a roulette wheel where each outcome is alloted an angle of $2\pi \mu_i$. We call the chosen value w_1. We then determine which of the three level 2 cells $F_{w_1} F_i K$ x belongs to by another, independent spin of the same roulette wheel, and so on.

One of the main reasons for wanting to have a measure is to be able to integrate functions. Since the functions we want to integrate are usually continuous, this is easily accomplished by imitating the ordinary integral in calculus: We subdivide the space into a union of essentially disjoint small sets $\{A_j\}$ and take the limit of sums $\sum_j f(x_j) \mu(A_j)$, where $x_j \in A_j$. Since f is continuous, hence uniformly continuous on the compact space K, the choice of the point $x_j \in A_j$ does not matter in the limit.

In our setup there is a natural choice of subdivisions, namely

$$(1.2.13) \qquad K = \bigcup_{|w|=m} F_w K,$$

the subdivision into all cells of level m. (See the exercises for other choices.) Then

$$(1.2.14) \qquad \int_K f d\mu = \lim_{m \to \infty} \sum_{|w|=m} f(x_w)\mu(F_w K)$$

for $x_w \in F_w K$. It is not difficult to show that the limit exists and satisfies the usual properties of integrals: linearity in f, and the estimate

$$(1.2.15) \qquad \min_K f \le \int_K f d\mu \le \max_K f.$$

If A is any finite union of cells, we can define $\int_A f d\mu$ by restricting to cells contained in A on the right side of (1.2.14). In analogy with the usual trapezoidal rule we may replace $f(x_w)$ by the average of f over the boundary points of the cell,

$$(1.2.16) \qquad \int_K f d\mu = \lim_{m \to \infty} \frac{1}{3} \sum_{i=0}^{2} \sum_{|w|=m} f(F_w q_i)\mu(F_w K),$$

in the case of SG. This has the advantage of exhibiting the integral as a limit of discrete graph integrals. Given a graph, if we assign probabilities $v(x)$ to the vertices, we write

$$(1.2.17) \qquad \int_\Gamma f dv = \sum_{x \in V} f(x)v(x).$$

Then (1.2.16) may be written

$$(1.2.18) \qquad \int_K f d\mu = \lim_{m \to \infty} \int_{\Gamma_m} f dv_m,$$

where $v_m(x)$ is defined to be

$$\frac{1}{3}(\mu(F_w K) + \mu(F_{w'} K)) \text{ if } x \in V_m \setminus V_0$$

has the addresses $x = F_w q_i = F_{w'} q_j$, and $\frac{1}{3}\mu(F_i^m K)$ if $x = q_i$. For the standard measure this is simply

$$(1.2.19) \qquad \int_{\Gamma_m} f dv_m = 3^{-m} \left(\frac{2}{3} \sum_{x \in V_m \setminus V_0} f(x) + \frac{1}{3} \sum_{x \in V_0} f(x) \right).$$

Note that we could drop the sum over the boundary points since this goes to zero in the limit. Nevertheless, for certain applications it is better to include them. The factor $\frac{2}{3}$ on the right side of (1.2.19) will play a role in the pointwise formula for the Laplacian.

In the case of the interval I, if we use the standard measure then we get the usual integral. For other choices of measure we get different integrals. There is a theorem that says $\int_I f d\mu = \int_0^1 f \circ \psi(x) dx$ for a suitable choice of a continuous change of variable function ψ (depending only on μ), but this function will be very irregular, certainly not differentiable.

If we fix a positive function f such that $\int_K f d\mu = 1$, then

(1.2.20) $$\nu(A) = \int_A f d\mu$$

defines another measure. The measure ν is called *absolutely continuous* with respect to μ. In fact, the correct definition requires that we allow a much broader class of functions, including discontinuous and unbounded functions. You might wonder if it is possible to pass from one self-similar measure to another by such a construction, but in fact it is not possible. It is easy to see that it is impossible using a bounded function f, since there are many cells where the ratio $\mu'(C)/\mu(C)$ is larger than any fixed constant, if μ and μ' are distinct self-similar measures.

We observe that it is possible to transform the self-similar identity (1.2.12) into an identity involving integrals of functions. Indeed, if $f = \chi_A$, the characteristic function of the set A (not really continuous, but having only a finite set of discontinuities, so that the integral may be defined as before), then

(1.2.21) $$\int_K f d\mu = \sum_i \mu_i \int_K f \circ F_i d\mu$$

is the same as (1.2.12) (note that $f \circ F_i = \chi_{F_i^{-1}A}$). By taking linear combinations and passing to the limit, it follows that (1.2.21) holds for all continuous functions.

EXERCISES

1.2.1. Show that the self-similar identity (1.2.12) generalizes to

$$\mu(A) = \sum_{|w|=m} \mu_w \mu(F_w^{-1}A) \quad \text{for any } m,$$

and similarly (1.2.21) generalizes to

$$\int_K f d\mu = \sum_{|w|=m} \mu_w \int_K f \circ F_w d\mu.$$

1.2.2. Let $\mathcal{P}$ be a finite set of words such that

$$K = \bigcup_{w \in \mathcal{P}} F_w K,$$

disjoint except for point intersections. We call $\mathcal{P}$ a *partition*. Show that

$$\mu(A) = \sum_{w \in \mathcal{P}} \mu_w \mu(F_w^{-1}A)$$

and

$$\int_K f d\mu = \sum_{w \in \mathcal{P}} \mu_w \int f \circ F_w d\mu.$$

1.2.3. Let $\rho = \min \mu_j$. Show that for any given $r > 0$ there exists a partition $\mathcal{P}$ such that $\rho r \leq \mu_w \leq r$ for every $w \in \mathcal{P}$.

1.2.4. Suppose $\mu_i \neq \mu_j$ for some $i \neq j$. Show that there exist adjacent cells $F_w K$ and $F_{w'} K$ with $|w| = |w'|$ such that $\mu(F_w K)/\mu(F_{w'} K)$ is as close to zero as desired.

1.2.5. Let μ be a self-similar measure on I. Use (1.2.21) to compute $\int_I x \, d\mu(x)$. Do the same for $\int_I x^2 d\mu(x)$.

1.3 GRAPH ENERGIES

Given a finite, connected graph G and a real-valued function u on its vertices, we define the graph energy by

$$(1.3.1) \qquad E_G(u) = \sum_{x \sim y} (u(x) - u(y))^2.$$

Here the sum extends over all edges of the graph. If we were to sum first over all vertices x and then over all neighboring vertices, then each edge would occur twice and we would compensate by multiplying by a factor of $\frac{1}{2}$. Energy is a quadratic form in u. We will also need the associated bilinear form

$$(1.3.2) \qquad E_G(u, v) = \sum_{x \sim y} (u(x) - u(y))(v(x) - v(y))$$

for pairs of functions. Of course $E_G(u) = E_G(u, u)$, and we can recover the bilinear form from the quadratic form by the usual polarization identity:

$$(1.3.3) \qquad E_G(u, v) = \frac{1}{4}(E_G(u + v) - E_G(u - v)).$$

It is clear that $E_G(u) = 0$ if u is constant, and the converse holds since we are assuming that G is connected. Also, $E_G(u, v)$ is an inner product on the space of functions on V modulo constants.

Another property of energy is called the *Markov property*: If we replace u by the minimum (or maximum) of u and a constant, the energy cannot increase. The reason for this is simply that each term $(u(x) - u(y))^2$ either stays the same or decreases. This is often stated in the form $E_G([u]) \leq E_G(u)$ for $[u] = \min\{1, \max\{u, 0\}\}$.

Now suppose we have two graphs, G and G', such that $V \subseteq V'$. We will think of G as a subgraph of G'. (We do not make any assumptions concerning the edges of G and G'.) Given a function u' on V', we can always restrict it to get a function $u = u'|_V$ on V. There is no apparent relationship between the energies $E_{G'}(u')$ and $E_G(u)$. If we go the other direction, starting with u defined on V, there are many possible extensions to V'. It is clear that there is at least one extension that minimizes the energy $E_{G'}(u')$. We will write $\tilde{u}$ for such an energy-minimizing extension and call it a *harmonic extension* (in the examples of interest, there will be a unique harmonic extension): $\tilde{u}|_V = u$ and $E_{G'}(\tilde{u}) \leq E_{G'}(u')$ for all u' such that $u'|_V = u$. We will call $E_{G'}(\tilde{u})$ the *restriction* of $E_{G'}$ to G.

Now it might happen, if we are lucky, that the restriction of $E_{G'}$ to G is equal to a multiple of E_G,

$$(1.3.4) \qquad E_{G'}(\tilde{u}) = r E_G(u),$$

for all functions u on V. We call (1.3.4) a *renormalization* equation. Typically, $0 < r < 1$. This means that if we renormalize the definition of energy on G' by multiplying by $1/r$, then the restriction to G gives the same value, $r^{-1} E_{G'}(\tilde{u}) = E_G(u)$, and since $\tilde{u}$ is an energy minimizer, we have

$$(1.3.5) \qquad\qquad r^{-1} E_{G'}(u') \geq E_G(u)$$

for every extension u' of u. In other words, energy increases with extension, except in the case of harmonic extension, when it remains unchanged.

This might seem like wishful thinking, but it actually works for the sequences of graphs Γ_m approximating K in both cases, I and SG! Let's look at I first. The first graph Γ_0 just consists of the vertices $\{0, 1\}$ connected by an edge, while Γ_1 consists of three vertices $\{0, \frac{1}{2}, 1\}$ connected sequentially. So the energies are given explicitly by (we change notation for simplicity)

$$(1.3.6) \qquad\qquad E_0(u) = (u(1) - u(0))^2$$

and

$$(1.3.7) \qquad E_1(u') = \left(u'(1) - u'\left(\frac{1}{2}\right) \right)^2 + \left(u'\left(\frac{1}{2}\right) - u'(0) \right)^2.$$

If u' is an extension of u, then $u'(1) = u(1)$ and $u'(0) = u(0)$, so the only issue is, What is $u'(\frac{1}{2})$? To minimize $E_1(u')$ and so obtain the harmonic extension, it seems obvious that we should take

$$(1.3.8) \qquad\qquad \tilde{u}\left(\frac{1}{2}\right) = \frac{1}{2}(u(1) + u(0)),$$

the linear extension (if we set $u'(\frac{1}{2}) = x$ and find the value where the x-derivative vanishes, we obtain (1.3.8)). A simple computation then reveals that

$$(1.3.9) \qquad\qquad E_1(\tilde{u}) = \frac{1}{2} E_0(u),$$

a renormalization equation with $r = \frac{1}{2}$.

But now consider what happens when we go from Γ_m to Γ_{m+1}. The vertices V_{m+1} consist of all points of the form $\frac{k}{2^{m+1}}$, and among them, those with k even belong to V_m, while those with k odd are new. If u is defined on V_m, the question of harmonic extension is, What is $\tilde{u}\left(\frac{k}{2^{m+1}}\right)$ when k is odd? At first, minimizing energy may seem like a global problem, but in fact it is entirely local! Fix an odd value, say $k = 2j + 1$. Then $\tilde{u}\left(\frac{2j+1}{2^{m+1}}\right)$ only appears twice in $E_{m+1}(\tilde{u})$, specifically in the terms

$$(1.3.10) \qquad \left(u\left(\frac{2j+2}{2^{m+1}}\right) - \tilde{u}\left(\frac{2j+1}{2^{m+1}}\right) \right)^2 + \left(\tilde{u}\left(\frac{2j+1}{2^{m+1}}\right) - u\left(\frac{2j}{2^{m+1}}\right) \right)^2.$$

This is the identical problem to the minimization of (1.3.7), and has the identical solution: Interpolate linearly,

$$(1.3.11) \qquad \tilde{u}\left(\frac{2j+1}{2^{m+1}}\right) = \frac{1}{2}\left(u\left(\frac{2j+2}{2^{m+1}}\right) + u\left(\frac{2j}{2^{m+1}}\right)\right).$$

Then the same computation that yielded (1.3.9) shows that (1.3.10) is equal to $\frac{1}{2}\left(u\left(\frac{2j+2}{2^{m+1}}\right) - u\left(\frac{2j}{2^{m+1}}\right)\right)^2$, and summing over j we obtain

$$(1.3.12) \qquad E_{m+1}(\tilde{u}) = \frac{1}{2}E_m(u),$$

a renormalization equation with the same constant $r = \frac{1}{2}$.

We define the renormalized graph energies by

$$(1.3.13) \qquad \mathcal{E}_m(u) = r^{-m}E_m(u),$$

for $r = 1/2$. For any function, $\{\mathcal{E}_m(u)\}$ is a nondecreasing sequence. It is in fact constant when u is a linear function. A linear function (at least on the set V_* of dyadic rationals) is uniquely determined by its boundary values $u(0)$ and $u(1)$ by repeated use of the local extension algorithm (1.3.11). This may seem banal, because linear functions are such easily understood objects, but it will help us to understand the less trivial analog on SG.

The renormalized energy may be written explicitly as

$$2^m \sum_{k=1}^{2^m}\left(u\left(\frac{k}{2^m}\right) - u\left(\frac{k-1}{2^m}\right)\right)^2 = \sum_{k=1}^{2^m}\left(\frac{u\left(\frac{k}{2^m}\right) - u\left(\frac{k-1}{2^m}\right)}{\frac{1}{2^m}}\right)^2\frac{1}{2^m}.$$

If u is continuously differentiable, the mean value theorem allows us to write this as

$$\sum_{k=1}^{2^m}(u'(x_k))^2\frac{1}{2^m}$$

for $\frac{k-1}{2^m} \le x_k \le \frac{k}{2^m}$, a Riemann sum for the integral

$$(1.3.14) \qquad \int_0^1 u'(x)^2dx.$$

So in that case $\mathcal{E}_m(u)$ converges to (1.3.14). We can also look at the renormalized bilinear form $\mathcal{E}_m(u, v) = r^{-m}E_m(u, v)$ and see that

$$\lim_{m\to\infty}\mathcal{E}_m(u, v) = \int_0^1 u'(x)v'(x)dx$$

if u and v are both continuously differentiable. We already know that if u is linear then $\mathcal{E}_m(u)$ is constant, but we may also assert that $\mathcal{E}_m(u, v)$ is constant for any function v. Indeed $u'(x)$ is then a constant, namely $u(1) - u(0)$, so

$$\int_0^1 u'(x)v'(x)dx = (u(1) - u(0))\int_0^1 v'(x)dx = E_0(u, v),$$

and by splitting the integral at the points $k/2^m$ we also obtain $\int_0^1 u'(x)v'(x)dx = \mathcal{E}_m(u, v)$. We may also observe directly that

(1.3.15)
$$\begin{aligned}
\mathcal{E}_1(u, v) &= 2\left(u(1) - u\left(\frac{1}{2}\right)\right)\left(v(1) - v\left(\frac{1}{2}\right)\right) \\
&\quad + 2\left(u\left(\frac{1}{2}\right) - u(0)\right)\left(v\left(\frac{1}{2}\right) - v(0)\right) \\
&= (u(1) - u(0))\left[\left(v(1) - v\left(\frac{1}{2}\right)\right) + \left(v\left(\frac{1}{2}\right) - v(0)\right)\right] \\
&= \mathcal{E}_0(u, v)
\end{aligned}$$

since $u(1) - u(\frac{1}{2}) = u(\frac{1}{2}) - u(0) = \frac{1}{2}(u(1) - u(0))$, etc.

Next we consider the case of SG. To keep the computation as simple as possible, we will exploit the symmetry. Suppose u is defined on V_0 by $u(q_0) = 1$ and $u(q_1) = u(q_2) = 0$, so $\mathcal{E}_0(u) = 2$, and we want to extend u to V_1 to minimize energy. By symmetry we will have $\tilde{u}(F_0 q_1) = \tilde{u}(F_0 q_2) = x$ at the junction points near q_0 and $\tilde{u}(F_1 q_2) = y$ at the junction point opposite q_0 in V_1, where x and y are to be determined (see Figure 1.3.1). Then

(1.3.16) $$E_1(\tilde{u}) = 2(x - 1)^2 + 2x^2 + 2y^2 + 2(x - y)^2$$

is to be minimized. By calculus we set the x and y derivatives equal to zero, to obtain the pair of linear equations

(1.3.17) $$\begin{cases} 4x = 1 + x + y, \\ 4y = 2x. \end{cases}$$

Note that these equations express the mean value property that the function value at each of the junction points is the average of the function values of the four neighboring points in the graph. The solution $x = \frac{2}{5}$, $y = \frac{1}{5}$ is clear by inspection. By symmetric we would get the same answer if we put the value 1 at any of the boundary vertices. Also, since we are minimizing a quadratic function, the minimizing equations are linear. So if the initial values of u on V_0 are a, b, c, then the harmonic extension $\tilde{u}$ satisfies the following "$\frac{1}{5} - \frac{2}{5}$ rule":

(1.3.18) $$u(z) = \frac{2}{5}a + \frac{2}{5}b + \frac{1}{5}c$$

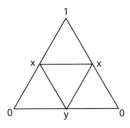

Figure 1.3.1 Values of $\tilde{u}$ on V_1.

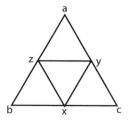

Figure 1.3.2 Values on V_1.

if z is the junction point between the vertices where u takes on the values a and b. Written this way, the harmonic extension will satisfy (1.3.18) on any cell of any level.

A more direct approach is to label the values on the vertices of V_1 as in Figure 1.3.2 and minimize E_1 as a function of x, y, z. The derivative equations yield the mean value equations

$$4x = b+c+y+z,$$
$$4y = a+c+x+z,$$
$$4z = a+b+x+y.$$

Adding these equations yields

$$x+y+z = a+b+c,$$

so

$$5x = b+c+(x+y+z) = b+c+(a+b+c),$$

and so on.

Finally, we need to compute the renormalization factor. For the function in Figure 1.3.1 with $x = \frac{2}{5}$ and $y = \frac{1}{5}$, we find

$$E_1(\tilde{u}) = 2\left(1 - \frac{2}{5}\right)^2 + 2\left(\frac{2}{5} - \frac{1}{5}\right)^2 + 2\left(\frac{2}{5} - 0\right)^2 + 2\left(\frac{1}{5} - 0\right)^2$$
$$= \frac{18 + 2 + 8 + 2}{25} = \frac{6}{5},$$

so the choice $r = \frac{3}{5}$ yields

(1.3.19) $$\mathcal{E}_1(\tilde{u}) = r^{-1}E_1(\tilde{u}) = E_0(u).$$

A little more work shows the same is true for the harmonic extension in the general case in Figure 1.3.2. Of course, a trivial remark is that the problem of minimizing the renormalized energy $\mathcal{E}_1$ is equivalent to minimizing E_1, with the same function $\tilde{u}$ achieving the minimum.

The same idea applies to the harmonic extension from V_1 to V_2, and in general from V_m to V_{m+1}. Suppose the values of u are given on V_m. Any new point in V_{m+1} (not in V_m) belongs to a unique m-cell $F_w K$ with $|w| = m$. The total energy $E_{m+1}(u')$ for any extension is simply the sum of contributions from each cell $F_w K$,

$$E_{m+1}(u') = \sum_{|w|=m} (u'(F_w F_0 q_0) - u'(F_w F_0 q_1))^2 + \cdots = \sum_{|w|=m} E_1(u' \circ F_w),$$

and each contribution is just the energy E_1 of $u' \circ F_w$. So the global minimization problem is just the union of 3^m local minimization problems of the sort we have just solved. So the "$\frac{1}{5} - \frac{2}{5}$ rule" (1.3.18) continues to hold on each m-cell for the harmonic extension, and the renormalization factor is again $r = \frac{3}{5}$. Altogether, if we define

$$(1.3.20) \qquad \mathcal{E}_m(u) = \left(\frac{3}{5}\right)^{-m} E_m(u),$$

then this renormalized energy remains unchanged under harmonic (minimum energy) extension, so it must go up for any extension:

$$(1.3.21) \qquad \mathcal{E}_0(u) \leq \mathcal{E}_1(u) \leq \mathcal{E}_2(u) \cdots .$$

In the next section we will take the limit of this sequence.

To summarize what we have found so far: Given a function u on V_m, the harmonic extension $\tilde{u}$ to V_{m+1} may be characterized in three ways:

  (i) it minimizes $\mathcal{E}_{m+1}(\tilde{u})$ at the value $\mathcal{E}_m(u)$;
  (ii) at each new point $x \in V_{m+1} \setminus V_m$, $\tilde{u}(x)$ is the average of the values at the four neighboring points in V_{m+1};
  (iii) it satisfies the "$\frac{1}{5} - \frac{2}{5}$ rule" at the new points in $V_{m+1} \setminus V_m$.

We may extend the equality in (i) to the bilinear form: If $\tilde{u}$, $\tilde{v}$ are the harmonic extensions of u and v, then

$$(1.3.22) \qquad \mathcal{E}_{m+1}(\tilde{u}, \tilde{v}) = \mathcal{E}_m(u, v)$$

by the polarization identity (1.3.3), since harmonic extension is a linear transformation (from (iii)). As in the case of I, we can say more.

LEMMA 1.3.1 *Let u, v be defined on V_m, let $\tilde{u}$ be the harmonic extension of u, and let v' be any extension of v to V_{m+1}. Then*

$$(1.3.23) \qquad \mathcal{E}_{m+1}(\tilde{u}, v') = \mathcal{E}_m(u, v).$$

Proof: Because of (1.3.22) it suffices to show $\mathcal{E}_{m+1}(\tilde{u}, v'') = 0$ for $v'' = v' - \tilde{v}$. Note that v'' vanishes on V_m. From the definition,

$$E_{m+1}(\tilde{u}, v'') = \sum_{\substack{x \sim y \\ m+1}} (\tilde{u}(x) - \tilde{u}(y))(v''(x) - v''(y)).$$

Now collect all the terms that contain $v''(x)$ for a fixed x. If $x \in V_m$ then $v''(x) = 0$, so these terms contribute 0. But if $x \in V_{m+1}$, then $v''(x)$ multiplies

$$(1.3.24) \qquad \sum_{\substack{y \sim x \\ m+1}} (\tilde{u}(x) - u(y)),$$

and this vanishes by the mean value condition (ii). So $E_{m+1}(\tilde{u}, v'') = 0$ and hence $\mathcal{E}_{m+1}(\tilde{u}, v'') = 0$. $\qquad\square$

Let's look at the "$\frac{1}{5} - \frac{2}{5}$" rule more closely. It says that the value at any inside point is a weighted average of the values of boundary points. The weight is higher at the boundary points closest to the inside point, as is to be expected. I don't know of any explanatory argument for the exact values of the weights; they come from the computation. In Section 1.5 we will give another derivation for the value $r = \frac{3}{5}$, but it will also be the result of a different computation.

We define a *harmonic* function h to be one that minimizes $\mathcal{E}_m$ at all levels for the given boundary values on V_0. In other words, with $h(q_0), h(q_1), h(q_2)$ given, we inductively find $h|_{V_{m+1}}$ from $h|_{V_m}$ using the "$\frac{1}{5} - \frac{2}{5}$ rule." This is a local extension algorithm: If we want to zoom in to great depth in a small neighborhood, it is not necessary to compute h on the whole gasket. Specifically, if we want to know the values of h to level $m + k$ on the cell $F_w K$ for $|w| = m$, we only have to compute h on the cells $F_{w_1} K, F_{w_1 w_2} K, F_{w_1 w_2 w_3} K, \ldots, F_w K$ and then compute the values of h in complete detail for k more levels, for a total of $m + 3^k$ steps, as compared to 3^{m+k} steps for computing h on the whole gasket.

The space of harmonic functions, denoted $\mathcal{H}_0$, is three-dimensional. A simple basis $\{h_0, h_1, h_2\}$ is obtained by taking $h_j(q_j) = 1$ and $h_j(q_k) = 0$ for $k \neq j$. Certain properties of harmonic functions follow easily from the extension algorithm. Although h is initially defined only on V_*, it is uniformly continuous and so extends to a continuous function on K. It also satisfies the maximum principle: The maximum and minimum are attained on the boundary (and only on the boundary if the function is not constant). In the next chapter we will show that harmonic functions are exactly the solutions of the differential equation $\Delta h = 0$.

The renormalized energies $\mathcal{E}_m(h)$ are the same for all m, in particular for $m = 0$, so

$$(1.3.25) \qquad \mathcal{E}_m(h) = (h(q_0) - h(q_1))^2 + (h(q_1) - h(q_2))^2 + (h(q_2) - h(q_0))^2.$$

In particular, $\mathcal{E}_m(h) > 0$ if h is nonconstant. Of course, if we start with h constant on V_0, then it remains constant on V_*, and it has zero energy by (1.3.25). In particular, $h_0 + h_1 + h_2 \equiv 1$.

It is convenient to represent the harmonic extension algorithm by a set of three matrices A_0, A_1, A_2 that describe how the boundary values change as we move from a cell of level m to its three subcells of level $m + 1$. That is,

$$(1.3.26) \qquad h|_{F_i V_0} = A_i h|_{V_0}$$

if we think of each set of h-values as a 3-vector, and more generally

$$(1.3.27) \qquad h|_{F_w F_i V_0} = A_i h|_{F_w V_0}.$$

Indeed, (1.3.27) is just (1.3.26) applied to the function $h \circ F_w$, which is also a harmonic function. It is easy to see that
(1.3.28)

$$A_0 = \begin{pmatrix} 1 & 0 & 0 \\ \frac{2}{5} & \frac{2}{5} & \frac{1}{5} \\ \frac{2}{5} & \frac{1}{5} & \frac{2}{5} \end{pmatrix}, \quad A_1 = \begin{pmatrix} \frac{2}{5} & \frac{2}{5} & \frac{1}{5} \\ 0 & 1 & 0 \\ \frac{1}{5} & \frac{2}{5} & \frac{2}{5} \end{pmatrix}, \quad A_2 = \begin{pmatrix} \frac{2}{5} & \frac{1}{5} & \frac{2}{5} \\ \frac{1}{5} & \frac{2}{5} & \frac{2}{5} \\ 0 & 0 & 1 \end{pmatrix}.$$

Another way of looking at it is that A_i is the matrix that represents the linear transformation $h \to h \circ F_i$ with respect to the basis $\{h_0, h_1, h_2\}$. Using the notation

$A_w = A_{w_m} \cdots A_{w_2} A_{w_1}$, we have

(1.3.29) $$h|_{F_w V_0} = A_w h|_{V_0}$$

(if you are wondering about the correct order in the product, work out the case $m = 2$). It is important to understand that this is all there is! Unlike the case of the interval, there is no other description of harmonic functions. In principle it should be possible to obtain any desired information about harmonic functions from (1.3.29). In practice this may require a lot of work!

The individual matrices A_i are easy to understand. Each has eigenvalues $1, \frac{3}{5}, \frac{1}{5}$. The eigenvector associated to 1 is the constant, but the eigenvectors associated to the other eigenvalues vary with the choice of i. For example, for A_0 the eigenvectors are $h_1 + h_2$ and $h_1 - h_2$ for eigenvalues $\frac{3}{5}$ and $\frac{1}{5}$, respectively. If we denote by R_0 the reflection symmetry that fixes q_0 and interchanges q_1 and q_2, then $h_1 + h_2$ is symmetric and $h_1 - h_2$ is skew-symmetric under R_0. If h is a harmonic function that vanishes at q_0, then it is a linear combination of $h_1 + h_2$ and $h_1 - h_2$ (write h as the sum of its symmetric and skew-symmetric parts). These functions have different decay rates as we approach q_0. Specifically, on the m-cell $F_0^m K$, $h_1 + h_2$ is $O((\frac{3}{5})^m)$ and $h_1 - h_2$ is $O((\frac{1}{5}))^m)$. A generic harmonic function vanishing at q_0 will have a nonzero symmetric part, so it will decay $O((\frac{3}{5})^m)$. To obtain the faster decay rate we have to choose a multiple of $h_1 - h_2$. In the next chapter we will see how to distinguish these cases by means of *normal derivatives*. The fact that the middle eigenvalue $\frac{3}{5}$ coincides with the renormalization constant r is no coincidence. The fact that $\frac{1}{5}$ is the smallest eigenvalue and 5 is the renormalization constant for the Laplacian is a coincidence. The numerology of these eigenvalues will have interesting consequences.

EXERCISES

1.3.1. The matrices A_i are invertible. Compute A_i^{-1} explicitly. Use these matrices to show how a harmonic function is uniquely determined by its values on the boundary of any given m-cell.

1.3.2. Consider the restriction of a harmonic function to the line segment in SG joining q_0 to q_1, and parametrize this segment by the unit interval in the obvious way. Find explicit formulas for $h(\frac{1}{4})$ and $h(\frac{3}{4})$ as a linear combination of $h(0), h(\frac{1}{2}), h(1)$. Show that this algorithm localizes, so the values of h on all dyadic points in the interval (vertex points in the segment) are determined by $h(0), h(\frac{1}{2}), h(1)$.

1.3.3.* Show that the restriction of h to this segment can have at most one local extremum.

1.3.4. Consider the two-dimensional space obtained from $\mathcal{H}$ by factoring out the constants. Choose a basis for this space and find explicit 2×2 matrices $\tilde{A}_i$ that represent the transformations $h \to h \circ F_i$ with respect to your basis. Note: There is no basis that is symmetric with respect to the dihedral group, so the result will not be as nice as (1.3.26), although each matrix $\tilde{A}_i$ will have eigenvalues $\frac{3}{5}$, $\frac{1}{5}$ and can be made symmetric if the basis is chosen appropriately.

1.3.5. Note that for a nonconstant harmonic function,

$$\mathcal{E}_1(h) = r^{-1} \sum_{i=0}^{2} \mathcal{E}_0(h \circ F_i)$$

gives a decomposition of the energy of h into parts of the energy $r^{-1}\mathcal{E}_0(h \circ F_i)$ coming from each of the cells $F_i K$. Show that it is impossible to find h for which all these values $r^{-1}\mathcal{E}_0(h \circ F_i)$ are equal. More generally, describe all possible ways that the energy can be split.

1.3.6. Let $\mathrm{Osc}(f, A)$ denote the difference between the maximum and minimum values of f on A. Show that for harmonic functions $\mathrm{Osc}(h, F_i K) \leq \frac{3}{5} \mathrm{Osc}(h, K)$, and more generally

$$\mathrm{Osc}(h, F_w K) \leq \left(\frac{3}{5}\right)^m \mathrm{Osc}(h, K) \quad \text{if } |w| = m.$$

Use this to deduce that h on V_* is uniformly continuous.

1.3.7. Show that the eigenvalues of A_w for $|w| = m$ are $1, \lambda_1, \lambda_2$ where $\lambda_1 \lambda_2 = (\frac{3}{25})^m$, and also

$$\left(\frac{1}{5}\right)^m \leq |\lambda_1| \leq |\lambda_2| \leq \left(\frac{3}{5}\right)^m.$$

(Hint: Use the results of Exercise 1.3.4. The symmetry of $\tilde{A}_i$ implies the upper bound.)

1.3.8. Show that $\mathcal{E}_m(u^2) \leq 4M^2 \mathcal{E}_m(u)$ if $|u| \leq M$. Prove a similar bound for $\mathcal{E}_m(uv)$ in terms of $\mathcal{E}_m(u)$, $\mathcal{E}_m(v)$ and upper bounds for u and v.

1.3.9.* Partition the edges of the graph Γ_m into three types, horizontal, slanting right, and slanting left, and similarly write $\mathcal{E}_m(h)$ as a sum of three "directional" energies $\mathcal{E}_m(h) = \mathcal{E}_m^{(1)}(h) + \mathcal{E}_m^{(2)}(h) + \mathcal{E}_m^{(3)}(h)$ by restricting the sum defining $\mathcal{E}_m$ to each type of edge. For harmonic functions h, show that $\mathcal{E}_m^{(1)}(h)$ converges to $\frac{1}{3}\mathcal{E}_0(h)$ as $m \to \infty$, for $i = 1, 2, 3$.

1.3.10. Show that

$$\begin{pmatrix} h_0 \circ F_i \\ h_1 \circ F_i \\ h_2 \circ F_i \end{pmatrix} = A_i^* \begin{pmatrix} h_0 \\ h_1 \\ h_2 \end{pmatrix}.$$

1.4 ENERGY

In the previous section we constructed (for $K = I$ or SG) a sequence of energies $\mathcal{E}_m$ on Γ_m such that $\mathcal{E}_m(u)$ is increasing (nondecreasing) for any function u defined on V_*. It makes sense to define

(1.4.1) $$\mathcal{E}(u) = \lim_{m \to \infty} \mathcal{E}_m(u),$$

allowing the value $+\infty$. Moreover, it is clear that $\mathcal{E}(u) = 0$ if and only if u is constant. We say $u \in \mathrm{dom}\,\mathcal{E}$ (u belongs to the domain of the energy) if and only if $\mathcal{E}(u) < \infty$. We also say that u has *finite energy*. The definition of energy only involves the values of u on V_*, and we would really like to think of u as a function on K. We will see later that if u has finite energy then it is uniformly continuous on V_*, hence it has a unique continuous extension to K. By the way, this is not true in Euclidean spaces or manifolds of dimension 2 or more, so the graph approximation method does not work in those contexts.

In addition to showing that dom $\mathcal{E} \subseteq C(K)$, we will show that dom $\mathcal{E}$ is dense in $C(K)$, so that there exists an adequate supply of functions of finite energy. It is clear from the previous section that harmonic functions have finite energy, and an easy extension of this idea is that piecewise harmonic functions (start with any values on $V_{m'}$ for some fixed m' and extend harmonically for $m > m'$) also have finite energy. In fact, we will show that piecewise harmonic functions are dense, both in $C(K)$ and in dom $\mathcal{E}$ in an appropriate sense.

Let u be a function of finite energy. Then $\mathcal{E}_m(u) \le \mathcal{E}(u)$, so if $x \underset{m}{\sim} y$ for $x, y \in V_m$ we have $r^{-m}(u(x) - u(y))^2 \le \mathcal{E}_m(u) \le \mathcal{E}(u)$ since $r^{-m}(u(x) - u(y))^2$ is a summand in $\mathcal{E}_m(u)$. This means

$$(1.4.2) \qquad\qquad |u(x) - u(y)| \le r^{m/2} \mathcal{E}(u)^{1/2}.$$

This is already a statement of continuity. Now consider a chain of points x_m, $x_{m+1}, \ldots, x_{m+k}$ such that $x_{m+j} \in V_{m+j}$ and $x_{m+j} \underset{m+j+1}{\sim} x_{m+j+1}$. Then we have

$$|u(x_m) - u(x_{m+k})| \le r^{m/2}(1 + r^{1/2} + \cdots + r^{k/2})\mathcal{E}(u)^{1/2} \le \frac{r^{m/2}}{1 - r^{1/2}}\mathcal{E}(u)^{1/2}$$

by adding up the estimates (1.4.2) along the chain of edges. From the geometry of K it is easy to see that if $x, y \in V_*$ belong to the same or adjacent m-cells, then we can connect x to y by at most two such chains, so

$$(1.4.3) \qquad\qquad |u(x) - u(y)| \le \frac{2r^{m/2}}{1 - r^{1/2}}\mathcal{E}(u)^{1/2}.$$

Not only is (1.4.3) a statement of uniform continuity, it is also a Hölder condition. In the case of the interval, if $|x - y| \le \frac{1}{2^m}$, then x and y belong to the same or adjacent m-cell. Since $r = \frac{1}{2}$, (1.4.3) says

$$(1.4.4) \qquad\qquad |u(x) - u(y)| \le M|x - y|^{1/2}.$$

(This is the optimal Hölder condition in the Sobolev embedding theorem for H^1, which may be identified with dom $\mathcal{E}$; see the exercises.) In the case of SG we also get a Hölder condition for the Euclidean metric with a strange exponent, $\log(\frac{5}{3})/\log 2$. In Section 1.6 we will introduce a more natural metric on SG, and with respect to this metric the Hölder exponent will again be $\frac{1}{2}$.

For the rest of this section, all functions will be assumed to be continuous and defined on all of K.

LEMMA 1.4.1 *Let* $u, v \in dom\, \mathcal{E}$. *Then*

$$(1.4.5) \qquad\qquad \lim_{m \to \infty} \mathcal{E}_m(u, v) = \mathcal{E}(u, v)$$

exists and defines an inner product on dom $\mathcal{E}/$ *constants.*

Proof: We begin with the polarization identity

$$(1.4.6) \qquad \mathcal{E}_m(u, v) = \frac{1}{4}(\mathcal{E}_m(u+v) - \mathcal{E}_m(u+v))$$

at level m. Since the right side of (1.4.6) has a limit, so does the left side. The usual properties of an inner product, except that $\mathcal{E}(u) = 0$ may occur, follow easily. Since $\mathcal{E}(u) = 0$ implies that $\mathcal{E}_m(u) = 0$, which implies that u is constant on V_m for all m, it follows that u must be constant. By factoring out by the constants, we obtain a true inner product. $\qquad\square$

THEOREM 1.4.2 *dom $\mathcal{E}$/constants forms a Hilbert space with inner product (1.4.5).*

Proof: It remains to show completeness: Every Cauchy sequence converges. It is convenient to identify dom $\mathcal{E}$/ constants with the space $\tilde{\mathcal{E}} = \{u \in \text{dom } \mathcal{E} : u(q_0) = 0\}$. Let $\{u_n\}$ be a sequence in $\tilde{\mathcal{E}}$ such that $\mathcal{E}(u_n - u_{n'}) \to 0$ as $n, n' \to \infty$. Then for fixed m, $\mathcal{E}_m(u_n - u_{n'}) \to 0$ also, since $\mathcal{E}_m(u_n - u_{n'}) \leq \mathcal{E}(u_n - u_{n'})$. It follows easily that

$$\lim_{n \to \infty} u_n(x) \quad \text{exists for each } x \in V_m,$$

so we may define u on V_* as this limit, and moreover

$$(1.4.7) \qquad \mathcal{E}_m(u_n - u) = \lim_{n' \to \infty} \mathcal{E}_m(u_n - u_{n'}).$$

By taking n large enough, the right side of (1.4.7) may be made as small as desired independent of m, so $\mathcal{E}(u_n - u) \to 0$ as $n \to \infty$. $\qquad\square$

Having to factor out by the constants is a minor nuisance. We will say $u_n \to u$ in energy if $\mathcal{E}(u_n - u) \to 0$ and also $u_n \to u$ uniformly (it suffices to have $u_n \to u$ at a single point in view of (1.4.2)).

DEFINITION 1.4.3 The space $S(\mathcal{H}_0, V_m)$ of piecewise harmonic splines of level m is defined to be the space of continuous functions such that $u \circ F_w$ is harmonic for all $|w| = m$.

It is easy to see that $S(\mathcal{H}_0, V_m)$ is contained in dom $\mathcal{E}$ and is a finite-dimensional space of dimension #V_m. All such functions are obtained by specifying the values of u on V_m arbitrarily and then extending harmonically to $V_{m'}$ for each $m' > m$. Clearly $\mathcal{E}(u) = \mathcal{E}_m(u)$ for these functions.

THEOREM 1.4.4 *Any function $u \in C(K)$ may be approximated uniformly by a sequence $u_m \in S(\mathcal{H}_0, V_m)$, with $u_m\big|_{V_m} = u\big|_{V_m}$. Moreover, if $u \in \text{dom } \mathcal{E}$ then u_m converges to u in energy.*

Proof: Given $\varepsilon > 0$, we can find m such that $\text{Osc}(u, F_w K) \leq \varepsilon$ for all w with $|w| = m$. Then since $u_m\big|_{V_m} = u\big|_{V_m}$ we also have $\text{Osc}(u_m, F_w K) \leq \varepsilon$, so

$$|u_m(x) - u(x)| \leq |u_m(x) - u_m(F_w q_0)| + |u_m(F_w q_0) - u(F_w q_0)|$$
$$+ |u(F_w(q_0) - u(x)|$$
$$\leq 2\varepsilon \text{ for } x \in F_w K,$$

so $\|u_m - u\|_\infty \leq 2\varepsilon$.

Next suppose $u \in \mathrm{dom}\, \mathcal{E}$. Then $\mathcal{E}_m(u) = \mathcal{E}_m(u_m) \nearrow \mathcal{E}(u)$. Also $\mathcal{E}(u, u_m) = \mathcal{E}_m(u, u_m) = \mathcal{E}_m(u_m)$ by Lemma 1.3.1. So

$$\mathcal{E}(u - u_m) = \mathcal{E}(u) - 2\mathcal{E}(u, u_m) + \mathcal{E}(u_m) = \mathcal{E}(u) - \mathcal{E}_m(u_m) \to 0.$$

□

The next result expresses the self-similarity of the energy.

THEOREM 1.4.5 *If $u \in \mathrm{dom}\,\mathcal{E}$ then $u \circ F_i \in \mathrm{dom}\,\mathcal{E}$ for all i, and*

$$(1.4.8) \qquad\qquad \mathcal{E}(u) = \sum_i r^{-1} \mathcal{E}(u \circ F_i).$$

Proof: It is clear from the definition that

$$\mathcal{E}_{m+1}(u) = \sum_i r^{-1} \mathcal{E}_m(u \circ F_i).$$

Taking the limit we obtain (1.4.8), and if the left side is finite then each term on the right must also be finite. □

Of course the same identity holds for the bilinear form $\mathcal{E}(u, v)$ and for sub-divisions

$$(1.4.9) \qquad\qquad K = \bigcup_{w \in \mathcal{P}} F_w K$$

with r^{-1} replaced by $r^{-|w|}$, for any partition $\mathcal{P}$:

$$(1.4.10) \qquad\qquad \mathcal{E}(u) = \sum_{w \in \mathcal{P}} r^{-|w|} \mathcal{E}(u \circ F_w).$$

Another way of saying this is that we can create a function of finite energy on K by gluing together finite energy functions on cells $F_w K$ provided the functions match at junction points.

The additivity in (1.4.10) suggests that we could think of energy as a measure. More precisely, define a measure ν_u by

$$(1.4.11) \qquad\qquad \nu_u(F_w K) = r^{-|w|} \mathcal{E}(u \circ F_w).$$

Equivalently, $\nu_u(F_w K)$ is obtained as a limit of

$$(1.4.12) \qquad\qquad \sum_{x \underset{m}{\sim} y} r^{-m} (u(x) - u(y))^2,$$

where the sum is restricted to those edges lying in $F_w K$. It is easy to check that all the conditions for a regular measure are satisfied except strict positivity (for a probability measure we would need $\mathcal{E}(u) = 1$). Then

$$(1.4.13) \qquad\qquad \mathcal{E}(u) = \nu_u(K) = \int_K 1\, d\nu_u.$$

An interesting difference beween I and SG is that on I, the energy measures are absolutely continuous with respect to the standard measure, but on SG they are not.

In fact they are singular—roughly speaking, they concentrate mass too much in neighborhoods of junction points. This result was first proved by Kusuoka. A hint of why this is so is given in Exercise 1.3.5.

Here are a couple of simple properties of energy: The Markov property

$$(1.4.14) \qquad \mathcal{E}([u]) \le \mathcal{E}(u) \text{ for } [u] = \min\{1, \max\{u, 0\}\}$$

follows from the corresponding property for $\mathcal{E}_m$. Also, dom $\mathcal{E}$ forms an algebra under pointwise multiplication. We leave the verification to the exercises.

EXERCISES

1.4.1. Show that $\nu_u(F_w K) \le r^{|w|}\mathcal{E}(u)$ and so the continuity condition (1.2.3) holds for ν_u.

1.4.2. Show that dom $\mathcal{E}$ is an algebra, and find an estimate for $\mathcal{E}(uv)$ in terms of $\mathcal{E}(u)$, $\mathcal{E}(v)$, $\|u\|_\infty$, and $\|v\|_\infty$.

1.4.3. Let R be one of the reflection symmetries in D_3, and suppose $u, v \in$ dom $\mathcal{E}$ are, respectively, symmetric and skew-symmetric with respect to R. Show that $\mathcal{E}(u, v) = 0$.

1.4.4. On SG choose an orthornormal basis $\{h_1, h_2\}$ for $\mathcal{H}_0$/constants with respect to the energy inner product, and define the *Kusuoka measure* $\nu = \nu_{h_1} + \nu_{h_2}$. Show that this measure is independent of the choice of orthonormal basis.

1.4.5. Show that if $u \in C^1(I)$ then $u \in$ dom $\mathcal{E}$ and $\mathcal{E}(u) = \int_0^1 (u'(x))^2 dx$.

1.4.6.* On I, show that dom $\mathcal{E}$ may be identified with the Sobolev space H^1, with $\mathcal{E}(u) = \int_0^1 (u'(x))^2 dx$, where now u' is the distributional derivative and the integral is a Lebesgue integral.

1.4.7. Consider the skew-symmetric function u on SG defined by $u(F_0^k q_1) = 3^k$, $u(F_0^k q_2) = -3^k$, and extended to be harmonic on every cell not containing q_0 (see Figure 1.4.1). Show that u has infinite energy, but u is harmonic in the complement of q_0 (u satisfies the mean value condition at level m for any vertex in $V_m \setminus V_0$ not adjacent to q_0).

1.4.8. Show that if $u \in S(\mathcal{H}_0, V_m)$ and $v \in$ dom $\mathcal{E}$ then $\mathcal{E}(u, v) = \mathcal{E}_m(u, v)$.

1.4.9. (a) Show that the energy is *local*, meaning $u \cdot v \equiv 0$ implies $\mathcal{E}(u, v) = 0$.
(b) Show that the energy is *strongly local*, meaning that $\mathcal{E}(u, v) = 0$ if v is constant on the support of u.

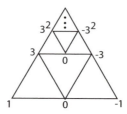

Figure 1.4.1

1.5 ELECTRIC NETWORK INTERPRETATION

In Section 1.3 we began by considering a notion of energy on a general finite graph G. More generally, suppose we have a positive function c_{xy} defined on the edges $x \sim y$ of the graph (we call this a *network*). Then we may consider

$$(1.5.1) \qquad E(u) = \sum_{x \sim y} c_{xy}(u(x) - u(y))^2$$

as an associated energy. We will interpret c_{xy} as conductances and the reciprocals $r_{xy} = 1/c_{xy}$ as resistances. We imagine an electric network where each vertex of G is a node and the edges of G are resistors connecting the nodes with the given resistances. The values of u are interpreted as voltages at the nodes. A current of amperage $(u(x) - u(y))/r_{xy} = c_{xy}(u(x) - u(y))$ will flow through each resistor, producing an energy of $c_{xy}(u(x) - u(y))^2$ from each resistor, leading to the total energy (1.5.1). Note that we have to do something (such as attach appropriate strength batteries) to keep the nodes at the specified voltages $u(x)$.

We could also drop the reference to the graph structure of G and require $c_{xy} \geq 0$ to be given as a symmetric function on all distinct pairs x, y in V. We then define $x \sim y$ if and only if $c_{xy} > 0$. If $c_{xy} = 0$ then the resistance is infinite, and it won't change the network to connect x and y by an infinite resistor. When we define the restriction of a network in what follows, we are essentially using this approach to define the edge relation.

We might also consider what happens if we impose voltages $u(x)$ at only some of the nodes (V') and allow the voltages at the other nodes (V'') to settle into values that, according to electric network theory, will minimize the energy. For example, suppose the network has three nodes x, y, z and two edges $x \sim y$ and $y \sim z$. If we set voltages $u(x)$ and $u(z)$ at the extreme nodes, then the value $u(y)$ at the middle node that minimizes

$$(1.5.2) \qquad c_{xy}(u(x) - u(y))^2 + c_{yz}(u(y) - u(z))^2$$

is easily seen to be

$$(1.5.3) \qquad u(y) = \frac{c_{xy}u(x) + c_{yz}u(z)}{c_{xy} + c_{yz}},$$

and this yields the value

$$(1.5.4) \qquad \left(\frac{c_{xy}c_{yz}}{c_{xy} + c_{yz}}\right)(u(x) - u(z))^2 = \frac{1}{r_{xy} + r_{yz}}(u(x) - u(z))^2$$

for (1.5.2). Note that this is the same value as the energy for a network with two nodes x and z connected by a resistor of resistance $r_{xy} + r_{yz}$. This is a familiar rule: Resistors in series add their resistances. The rule that resistors in parallel add their conductances is more or less built into the energy formula (1.5.1).

It seems reasonable that whatever the choice of nodes V', we could construct a network on V' that mimics the energy on the original network for any choice of values $u(x)$ for $x \in V'$. Any such network will be called a *restriction* of the original network, and the original network will be called an *extension* of the network on V'. We will not be concerned here with abstract existence and uniqueness theorems for restrictions, since in all cases of interest we will compute restrictions explicitly.

From this point of view, the solution of the renormalization problem relating the energies $\mathcal{E}_0$ and $\mathcal{E}_1$ on the graphs Γ_0 and Γ_1 for SG is the same as starting with a network on Γ_1 with all resistance equal to r and showing that the restriction to V_0 is the graph Γ_0 with all resistances equal to 1. Here we will re-derive the answer in a step-by-step fashion using four basic principles of electric network theory, applied to pieces of the graphs. We have already mentioned the first two, resistors in series and resistors in parallel. The others are "pruning" and the "$\Delta - Y$ transformation."

LEMMA 1.5.1 *Suppose the deleted vertices V'' are connected only to each other and to a single vertex x_0 in V'. Then the restriction network is obtained by retaining all the edges connecting nodes in V' with the same resistances, and the minimum energy function has $u(y) = u(x_0)$ for every $y \in V''$.*

Proof: Given u on V', the choice $u(y) = u(x_0)$ for all $y \in V''$ adds zero to the sum

$$\sum_{\substack{x \sim y \\ x, y \in V'}} c_{xy}(u(x) - u(y))^2$$

and so clearly minimizes energy. □

LEMMA 1.5.2 *Consider a Y-shaped network with nodes x, y, z, w and edges just connecting x, y, z to w, and resistances r_{xw}, r_{yw}, r_{zw}. Then the restriction to $V' = \{x, y, z\}$ is a Δ-shaped network with resistances r_{xy}, r_{yz}, r_{zx} provided that*

$$(1.5.5) \quad \begin{cases} r_{xw} = \frac{r_{xy} r_{zx}}{R}, & r_{yw} = \frac{r_{xy} r_{yz}}{R}, & r_{zw} = \frac{r_{yz} r_{zx}}{R} \\ \text{for} \quad R = r_{xy} + r_{yz} + r_{zx}. \end{cases}$$

Moreover, the energy-minimizing value is

$$(1.5.6) \quad u(w) = \frac{c_{xw} u(x) + c_{yw} u(y) + c_{zw} u(z)}{c_{xw} + c_{yw} + c_{zw}}.$$

In particular, if $r_{xy} = r_{yz} = r_{zx} = a$ then $r_{xw} = r_{yw} = r_{zw} = a/3$, and $u(w) = \frac{1}{3}(u(x) + u(y) + u(z))$.

Proof: The Y-network energy is

$$(1.5.7) \quad c_{xw}(u(x) - u(w))^2 + c_{yw}(u(y) - u(w))^2 + c_{zw}(u(z) - u(w))^2.$$

It is clear that to minimize this we must choose $u(w)$ by (1.5.6). When we substitute (1.5.6) into (1.5.7) and simplify we obtain

$$(1.5.8) \quad c_{xy}(u(x) - u(y))^2 + c_{yz}(u(y) - u(z))^2 + c_{zx}(u(z) - u(x))^2$$

for certain coefficients. After some messy algebraic manipulations we obtain (1.5.5). The details are left to the exercises. The special case of equal resistances is easy. □

Now we analyze the restriction of the Γ_2-network for SG with equal resistances. To simplify the computation we set the resistances equal to 1, since the result is

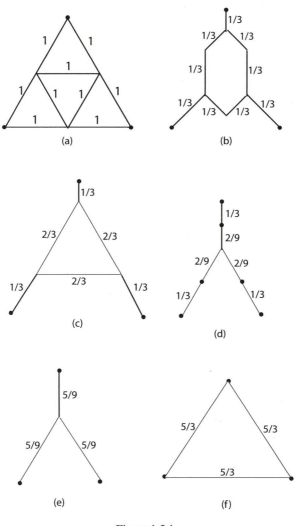

Figure 1.5.1

clearly linear in the resistance. We obtain the network in Figure 1.5.1(a), with the V_0 vertices marked by larger dots. We observe three Δ-shaped subnetworks, so we apply the $\Delta - Y$ transform to each, to obtain Figure 1.5.1(b). We see three sets of resistors in series, so we combine them to obtain Figure 1.5.1(c). Another Δ-shaped subnetwork appears, so we use $\Delta - Y$ to obtain Figure 1.5.1(d). Again there are three sets of resistors in series, so we combine to obtain the Y-shaped network in Figure 1.5.1(e). Finally, we do the inverse of the $\Delta - Y$ transform to obtain the network on V_0 in Figure 1.5.1(f). Keeping track of the resistances along the way, we see that we have multiplied by $\frac{5}{3}$, so if we started with $r = \frac{3}{5}$ we would end up with resistance 1. This confirms our calculation of the energy renormalization factor.

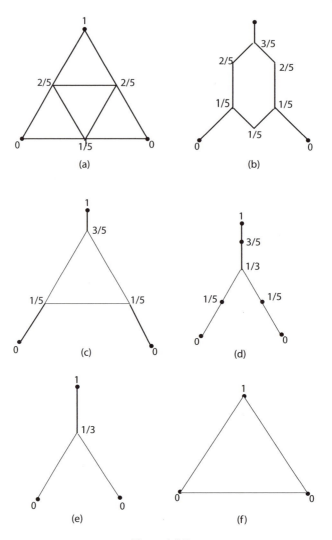

Figure 1.5.2

To obtain the harmonic extension algorithm we start with values of u at the vertices in Figure 1.5.1(f) and run the process backwards, using (1.5.3) and (1.5.6) to fill in the values when we add vertices. We show the results in Figure 1.5.2. For simplicity we just do the function h_0, but by symmetry the analogous computation holds for h_1 and h_2, and by linearity for all harmonic functions. Thus we rediscover the "$\frac{1}{5} - \frac{2}{5}$ rule".

Which method is easier? The first method involves solving a system of linear equations. The second method gives a step-by-step procedure that leads to the solution. Of course, Gaussian elimination also gives a step-by-step procedure that leads to the solution, but it is not the same procedure. In this case the system of equations was easy to solve, so the first method was simpler. In other examples it seems that network manipulation is simpler. But not all networks allow simiplification using the few rules we have at hand. So, from a practical point of view, network manipulation is a tool to be used sparingly, though sometimes to great effect. From a theoretical point of view, it motivates and explicates the notion of effective resistance metric that we discuss in the next section.

EXERCISES

1.5.1.* Show that the restriction of a network to a subset $V' \subseteq V$ defines a unique network on V'.

1.5.2. Suppose $V''' \subset V' \subseteq V$. Given a network on V, show that its restriction to V''' is equal to the restriction to V''' of the restriction to V'.

1.5.3. Supply the details of the proof of Lemma 1.5.2.

1.5.4. Show that a network is connected if and only if the only functions of zero energy are the constants. Use this to show that the restriction of a connected network is connected.

1.5.5. Let $x_1 < x_2 < \cdots < x_n$ be points on the line, and define a network with $x_j \sim x_{j+1}$ and $r_{x_j x_{j+1}} = x_{j+1} - x_j$. Determine the restriction of this network to any subset of $\{x_j\}$.

1.5.6. Invert (1.5.5) to express r_{xy}, r_{yz}, r_{zx} in terms of r_{xw}, r_{yw}, r_{zw}.

1.6 EFFECTIVE RESISTANCE METRIC

Given any network, we can define the effective resistance $R(x, y)$ between any two points as the resistance between them when we restrict the network to just those two points. This is exactly the resistance we would measure if we attached a battery to the two points and measured the current flow. It should not be confused with the resistance of an edge connecting x and y. Such an edge need not exist! If we look at the definition of restriction of networks, we find the following formula:

$$(1.6.1) \qquad R(x, y)^{-1} = \min\{\mathcal{E}(u) : u(x) = 0 \quad \text{and} \quad u(y) = 1\}.$$

Another formulation is that $R(x, y)$ is the minimum value of R such that

$$(1.6.2) \qquad |u(x) - u(y)|^2 \le R\mathcal{E}(u) \quad \text{for all } u \in \text{dom}\,\mathcal{E}.$$

We note that the function achieving the minimum in (1.6.1) is the function that is harmonic in the complement of the points x and y. For example, if the network is Γ_m for $K = I$, then (assuming $x < y$) the function u is 0 on $[0, x]$, 1 on $[y, 1]$, and linear on $[x, y]$. Clearly

$$\mathcal{E}_m(u) = \mathcal{E}(u) = \int_x^y \frac{1}{(y - x)^2}\, dt = \frac{1}{y - x},$$

so $R(x, y) = y - x$, the usual distance on I.

This leads us to hope that effective resistance will provide us with a natural (or intrinsic) metric on SG. What do we mean by this? The simplest interpretation is to first define $R(x, y)$ for points in V_m using the network Γ_m. Then we can extend it to V_*, since it clearly is independent of m (once m is large enough that $x, y \in V_m$). It is not difficult to see that $R(x, y)$ is uniformly continuous in x and y, so we may extend it to $SG \times SG$, and in fact (1.6.1) still holds. But is it a metric?

The claim is that effective resistance is a metric for any network. The only non-trivial condition to check is the triangle inequality. So given three points x, y, z, consider the restriction of the network to $\{x, y, z\}$. By Exercise 1.5.2, the effective resistances will be the same if we compute them with respect to this three-point network. So we only have to check the triangle inequality for a Δ-network. (There is also the trivial case when only two of the resistances are finite, where the triangle inequality is an equality.) This is an easy exercise, but it becomes quite obvious by doing a $\Delta - Y$ transformation. On the Y-network, $R(x, y) = r_{xw} + r_{yw}$, and so on, so

$$R(x, y) + R(y, z) = r_{xw} + r_{yw} + r_{yw} + r_{zw} > r_{xw} + r_{zw} = R(x, z).$$

Returning to SG, we know that effective resistance is a metric on V_m, hence on V_*, and by continuity on SG. We will soon see that it defines the same topology as the Euclidean metric, but it is not metrically equivalent. We note that (1.6.2), which may be written

$$(1.6.3) \qquad |u(x) - u(y)| \leq \mathcal{E}(u)^{1/2} R(x, y)^{1/2} \quad \text{for all } u \in \text{dom}\,\mathcal{E},$$

says that functions on $\text{dom}\,\mathcal{E}$ are Hölder continuous of order $\frac{1}{2}$ in the effective resistance metric.

It is extremely difficult to compute $R(x, y)$, but it is rather easy to obtain approximate values. First we note that if we can construct any function u satisfying $u(x) = 0$ and $u(y) = 1$, then this immediately gives us the lower bound

$$(1.6.4) \qquad\qquad\qquad R(x, y) \geq \mathcal{E}(u)^{-1}.$$

To find an upper bound we need to show that $u(x) = 0$ and $u(y) = 1$ implies $\mathcal{E}(u) \geq a$, as this implies

$$(1.6.5) \qquad\qquad\qquad R(x, y) \leq a^{-1}.$$

In particular, suppose $x, y \in V_m$ are neighboring vertices. Now we choose $u = \psi_y^{(m)}$, the piecewise harmonic spline in $S(\mathcal{H}_0, V_m)$ with $\psi_y^{(m)}(z) = \delta_{yz}$ for $y, z \in V_m$. Then we have $u(x) = 0$ and $u(y) = 1$, and $\mathcal{E}(u) = 4r^{-m}$ (or $2r^{-m}$ if y is a boundary point). So (1.6.4) says

$$(1.6.6) \qquad\qquad\qquad R(x, y) \geq \frac{1}{4} r^m.$$

On the other hand, $u(x) = 0$ and $u(y) = 1$ implies $\mathcal{E}(u) \geq \mathcal{E}_m(u) \geq r^{-m}$, so (1.6.5) says

$$(1.6.7) \qquad\qquad\qquad R(x, y) \leq r^m.$$

Together, the two estimates show that $R(x, y) \approx r^m$. The same reasoning extends to other points.

LEMMA 1.6.1 *There exist positive constants c_1 and c_2 such that*
(a) if x and y belong to the same or adjacent m-cells, then

$$(1.6.8) \qquad\qquad\qquad R(x, y) \le c_1 r^m;$$

(b) if x and y do not belong to the same or adjacent m-cells, then

$$(1.6.9) \qquad\qquad\qquad R(x, y) \ge c_2 r^m.$$

Proof: In case (a) we construct chains of points joining x and y as in the beginning of Section 1.4. Using the triangle inequality and estimate (1.6.7) for pairs of consecutive points in the chain, by summing a geometric series we obtain (1.6.8). In case (b), let z_0, z_1, z_2 denote the boundary points of an m-cell containing y. Then $u = \psi_{z_0}^{(m)} + \psi_{z_1}^{(m)} + \psi_{z_2}^{(m)}$ is identically 1 on the m-cell containing y, but $u(x) = 0$. Also $\mathcal{E}(u) = 6r^{-m}$ (or $4r^{-m}$ if the cell intersects V_0), so (1.6.4) implies (1.6.9). $\square$

It is easy to see that this means

$$(1.6.10) \qquad\qquad R(x, y) \sim |x - y|^{\beta} \quad \text{for } \beta = \log \frac{5}{3} / \log 2.$$

This shows that the resistance metric is topologically equivalent, but not metrically equivalent, to the Euclidean metric. Since $\beta < 1$, it follows that distances in the resistance metric are much larger than in the Euclidean metric. In particular, there are no rectifiable curves in this metric. It seems very unlikely that SG in this metric can be embedded in a Euclidean space of any dimension.

There is no exact scaling identity relating $R(x, y)$ and $R(F_i x, F_i y)$. We can say that, roughly speaking, each F_i acts like a contraction of ratio r.

EXERCISES

1.6.1. Show the equivalence of (1.6.1) and (1.6.2).

1.6.2. Prove that $R(x, y)$ is uniformly continuous on $V_* \times V_*$ in SG, and (1.6.1) holds on all of $SG \times SG$.

1.6.3. Compute the effective resistance on a Δ-network directly (without using the $\Delta - Y$ transform), and show that it is a metric.

1.6.4. Compute $R(x, y)$ exactly on $V_1 \times V_1$.

1.6.5. Give the details of the proof of Lemma 1.6.1(a).

1.6.6. Prove (1.6.10) from Lemma 1.6.1.

1.6.7. Show that $\mu\{x : R(x, y) \le r\} \sim r^d$ for $d = \frac{\log 3}{\log(5/3)}$, where μ is the standard measure. This means that SG as a metric-measure space, with metric R and measure μ, has dimension d.

1.7 NOTES AND REFERENCES

Most of the material in Sections 1.1, 1.3, and 1.4 is from [Kigami 1989], where it was developed for SG and its higher dimensional analogs. The fact that it also has something to say about I is a pleasant afterthought. Although it doesn't say anything new, it gives an amazingly simple characterization of the Sobolev space H^1; no Lebesgue integration theory or Schwartz distribution theory is needed, just a plain calculus-type limit. Of course, to show the equivalence of H^1 and dom $\mathcal{E}$ in Exercise 1.4.6 requires all this machinery.

The definition of self-similar measure in Section 1.2 is from [Hutchinson 1981]. We take a very naive approach to measures in general because we only have to integrate continuous or piecewise continuous functions, so we can use a Riemann-type integral. The positivity (1.2.1) and continuity (1.2.3) conditions are not part of the usual definition of measure, and we have made the ad hoc definition of "regular" to describe them. Occasionally we have to consider more general measures in the sequel.

The topological rigidity of SG (Exercise 1.1.6) was first noted in [Bandt and Retta 1992]. The fact that $\widetilde{SG}$ is not topologically rigid (Exercise 1.1.7) can also be understood from the Apollonian packing model. See [Mumford et al. 2002] for beautiful pictures of this.

Exercise 1.3.5 is a warmup for the singularity of energy measures [Kusuoka 1989]. Exercise 1.3.9 is from [Stanley et al. 2003]. The singular harmonic function in Exercise 1.4.7 was first noted in [Dalrymple et al. 1999].

The electric network ideas in Sections 1.5 and 1.6 come from [Kigami 1994a]. See [Doyle and Snell 1984] for the general theory of networks. See [Fukushima et al. 1994] for the general theory of Dirichlet forms.

It is possible to identify dom $\mathcal{E}$ with a certain Lipschitz-type function space determined by the embedding of SG in the plane, as shown in [Jonsson 1996]. This is one result that contradicts my assertion that the standard embedding of SG in the plane is irrelevant for our analytic theory. Nevertheless, it seems to be a kind of isolated result, since other natural function spaces on SG are unrelated to the embedding [Strichartz 2003b].

Chapter Two

Laplacian

2.1 WEAK FORMULATION

We are now in a position to define, in a few lines, a Laplacian Δ on both our self-similar spaces, I and SG, via the same weak formulation. For this we require only two ingredients: the bilinear energy $\mathcal{E}(u, v)$ and a regular probability measure μ. For the most part we will take μ to be the standard self-similar measure, but it is interesting to observe that we are free to make other choices. We will write Δ_μ to denote the dependence on the choice of measure and Δ if μ is the standard measure, in which case we call Δ the *standard* Laplacian. The idea behind the definition is the integration-by-parts formula. A student emerging from a standard calculus course is likely to remember integration-by-parts as just one in a bag of tricks for evaluating recalcitrant integrals; and yet, it turns out to be one of the most versatile tools in all of modern analysis.

Suppose v is a C^2 function on I that happens to vanish at the endpoints. Then

$$(2.1.1) \qquad \int_0^1 u''(x)v(x)dx = -\int_0^1 u'(x)v'(x)dx$$

is valid for $u \in C^2$, but actually also works conversely: If u is C^1 and there is a continuous f such that

$$(2.1.2) \qquad \int_0^1 f(x)v(x)dx = -\int_0^1 u'(x)v'(x)dx$$

for all such v, then $u \in C^2$ and $u'' = f$. Notice that we have avoided the messiness of boundary terms by assuming $v(0) = v(1) = 0$. Don't worry, we'll come back to the boundary terms in Section 2.3. For now we can rewrite (2.1.2) as

$$(2.1.3) \qquad \mathcal{E}(u, v) = -\int_0^1 f(x)v(x)dx \quad \text{for all } v \in \mathrm{dom}_0\mathcal{E}$$

(recall that the subscript 0 means exactly that v vanishes on the boundary). So integration-by-parts tells us that $u \in C^2$ and $u'' = f$ if and only if $u \in \mathrm{dom}\,\mathcal{E}$ and (2.1.3) holds. (A minor technical point: $u \in \mathrm{dom}\,\mathcal{E}$ is actually weaker than $u \in C^1$.) We leave the proof to the exercises.

The following definition makes sense for $K = I$ or SG.

DEFINITION 2.1.1 Let $u \in \mathrm{dom}\,\mathcal{E}$ and let f be continuous. Then $u \in \mathrm{dom}\,\Delta_\mu$ with $\Delta_\mu u = f$ if

$$(2.1.4) \qquad \mathcal{E}(u, v) = -\int_K fv d\mu \quad \text{for all } v \in \mathrm{dom}_0\mathcal{E}.$$

(Note that the positivity condition (1.2.1) is needed in order for (2.1.4) to uniquely determine f.) More generally, if we only assume $f \in L^2(d\mu)$ and (2.1.4) holds, then we say $u \in \mathrm{dom}_{L^2}\Delta_\mu$ and $\Delta_\mu u = f$.

This definition is the weak formulation. Later we will prove that there is an equivalent pointwise formulation. One could just as well start with the pointwise formulation as the definition and prove the above definition as a theorem. However, this does not work for the broader class $\mathrm{dom}_{L^2}\Delta_\mu$. (We will mainly use this in situations where f is piecewise continuous but has a few jump discontinuities.) Although the weak formulation may seem artificial at first, it actually proves to be more useful in the long run.

One source of confusion in the case $K = I$ is that (2.1.4) is an equation with two distinct measures: on the left side the energy is always the integral $\int_0^1 u'(x)v'(x)dx$ with respect to the standard measure, which may be different from the measure μ in the integral on the right side. (The actual meaning of $\Delta_\mu u = f$ in the case of $K = I$ and μ a general measure is that as a distribution, $\frac{d^2}{dx^2}u = f d\mu$.) Of course, when $K = SG$ there is no integral on the left side of (2.1.4).

In order for a definition to be interesting mathematically, not only does it have to be logically sound, but it has to hold in a sufficient number of cases. A priori, it is not clear that there are any nontrivial functions in $\mathrm{dom}\,\Delta_\mu$. It is clear that $0 \in \mathrm{dom}\,\Delta_\mu$ with $\Delta 0 = 0$, or more generally any constant function, and also that $\mathrm{dom}\,\Delta_\mu$ forms a vector space of functions. But beyond that (at least for SG) we will have to do some work. Eventually we will see that for every continuous function f there exists $u \in \mathrm{dom}\,\Delta_\mu$ such that $\Delta_\mu u = f$—this is about the best we could hope for. To begin, we will show that $\mathrm{dom}\,\Delta_\mu$ contains harmonic functions h, and $\Delta_\mu h = 0$. Note that this is true for all measures μ.

THEOREM 2.1.2 *If h is harmonic, then $h \in \mathrm{dom}\,\Delta_\mu$ and $\Delta_\mu h = 0$. Conversely, if $u \in \mathrm{dom}\,\Delta_\mu$ and $\Delta_\mu u = 0$ then u is harmonic.*

Proof: By Lemma 1.3.1, $\mathcal{E}_m(h, v)$ is independent of m, so $\mathcal{E}(h, v) = \mathcal{E}_0(h, v)$. But $\mathcal{E}_0(h, v) = 0$ because v vanishes on the boundary. This shows $\Delta_\mu h = 0$. For the converse, we make a special choice of v. Given m and a point $x \in V_m \setminus V_0$, let $\psi_x^{(m)}$ denote the piecewise harmonic spline in $S(\mathcal{H}_0, V_m)$ satisfying $\psi_x^{(m)}(y) = \delta_{xy}$ for $y \in V_m$. Note that $\psi_x^{(m)} \in \mathrm{dom}_0\mathcal{E}$ because $x \notin V_0$. Then $\mathcal{E}(u, \psi_x^{(m)}) = 0$ because $\Delta_\mu u = 0$. But, again by Lemma 1.3.1 (reversing the roles of u and v), we have $\mathcal{E}(u, \psi_x^{(m)}) = \mathcal{E}_m(u, \psi_x^{(m)})$. However, the equation $\mathcal{E}_m(u, \psi_x^{(m)}) = 0$ is exactly the condition

$$\sum_{y \underset{m}{\sim} x} (u(x) - u(y)) = 0$$

that $u\big|_{V_m}$ be harmonic. Since this is true for all m, u is harmonic. $\square$

It might seem plausible that we could easily create other functions in $\mathrm{dom}\,\Delta_\mu$ by passing to wider classes of functions related to harmonic functions. But this is not so simple. For example, piecewise harmonic splines are not in $\mathrm{dom}\,\Delta_\mu$, this is certainly clear for $K = I$. Another approach would be to try powers and polynomials of harmonic functions; while this works for $K = I$, it turns out for SG that

$u \in \mathrm{dom}_\mu \Delta$ implies that u^2 does *not* belong to $\mathrm{dom}_\mu \Delta$ (for the standard measure), as we will see in Section 2.7. So, for a while we will develop some of the properties of dom Δ_μ before we can show that there actually are lots of functions there.

The first property is the scaling property, which holds in the case that μ is a self-similar measure. So assume that (1.2.19) holds for some probabilities $\{\mu_i\}$. This is a scaling property for the measure. We also have the scaling property for the energy

$$\mathcal{E}(u, v) = \sum_i r^{-1}\mathcal{E}(u \circ F_i, v \circ F_i).$$

Substituting these into (2.1.4) we obtain

(2.1.5) $\qquad -\sum_i r^{-1}\mathcal{E}(u \circ F_i, v \circ F_i) = \sum_i \mu_i \int_K (f \circ F_i)(v \circ F_i)d\mu.$

Now given a value $i = j$ and $w \in \mathrm{dom}_0\mathcal{E}$, we construct $v \in \mathrm{dom}_0\mathcal{E}$ such that $v \circ F_j = w$ and $v \circ F_i = 0$ for all $i \neq j$. Using this v in (2.1.5) yields

(2.1.6) $\qquad -r^{-1}\mathcal{E}(u \circ F_j, w) = \mu_j \int f \circ F_j w d\mu,$

so according to the defintion, $u \circ F_i \in \mathrm{dom}\,\Delta_\mu$ and

(2.1.7) $\qquad \Delta_\mu(u \circ F_j) = r\mu_j(\Delta_\mu u) \circ F_j.$

By iteration we obtain

(2.1.8) $\qquad \Delta_\mu(u \circ F_w) = r_w\mu_w(\Delta_\mu u) \circ F_w$

where $r_w = r^{|w|}$. In particular, for the standard measure the scaling factor $r\mu$ for the Laplacian is $\frac{1}{4}$ for $K = I$ and $\frac{1}{5}$ for $K = SG$.

The appearance of the scaling factor $\frac{1}{5}$ in the Laplacian on SG strikes many people as strange. The explanation we have given is that it is not an irreducible quantity, but rather the product of two factors, $\frac{3}{5}$ for scaling energy and $\frac{1}{3}$ for scaling measure. Nevertheless, it still is strange. The scaling factor for the Laplacian in any dimensional Euclidean space $\mathbb{R}^n$ is always $\frac{1}{4}$. So the initial expectation is that SG is a space of some fractional dimension, but the dimension shouldn't matter. A more sophisticated analysis might note that the contractions F_i have scaling factor $\frac{1}{2}$ with respect to the Euclidean metric, but essentially $\frac{3}{5}$ with respect to the resistance metric. This leads to an expected value of $(\frac{3}{5})^2$ for scaling factor for the Laplacian, still not correct. The only way out of this quandary is to give up the idea that the Laplacian is a second-order operator. Later we will see more evidence for this point of view.

EXERCISES

2.1.1. Suppose $u \in C^1(I)$ and f is continuous on I, and suppose (2.1.2) holds for every $v \in C^1$ with $v(0) = v(1) = 0$. Prove that $u \in C^2$ and $u'' = f$.

2.1.2.* Redo Exercise 2.1.1, but only assuming that u and v belong to dom $\mathcal{E}$.

2.1.3.* On I, show that $u \in \text{dom}\, \Delta_\mu$ with $\Delta_\mu u = f$ if and only if $\frac{d^2}{dx^2}u = f d\mu$ in the sense of distributions.

2.1.4.* On I, show that $\text{dom}_{L^2}\Delta_\mu$ for the standard measure μ may be identified with the Sobolev space H^2.

2.1.5.* Define $u \in \text{dom}_\mathcal{M}\Delta$ and $\Delta u = v$, where v is a finite signed measure with no atoms on ∂K, if $u \in \text{dom}\, \mathcal{E}$ and

$$-\mathcal{E}(u, v) = \int_K v\, dv \quad \text{for every } v \in \text{dom}_0 \mathcal{E}.$$

Show that if $u \in \text{dom}\, \Delta_\mu$ with $\Delta_\mu u = f$, then $u \in \text{dom}_\mathcal{M}\Delta$ with $\Delta u = f d\mu$. (Note: The drawback to this definition is that Δ maps functions to measures, so we can't take $\Delta^2 u$.)

2.1.6. Show that dom Δ_μ is a vector space, and $\Delta_\mu(c_1 u_1 + c_2 u_2) = c_1 \Delta_\mu u_1 + c_2 \Delta_\mu u_2$.

2.1.7. On SG, show that the space of solutions to $(\Delta_\mu)^2 u = 0$ has dimension at most 6.

2.2 POINTWISE FORMULA

But what is $\Delta_\mu u(x)$ at a particular point x? We would like an answer that is reminiscent of the second difference quotient formula

$$(2.2.1) \qquad u''(x) = \lim_{m \to \infty} \frac{u\left(x + \frac{1}{2^m}\right) - 2u(x) + u\left(x - \frac{1}{2^m}\right)}{\left(\frac{1}{2^m}\right)^2}$$

on the interval. Another way of looking at it is that there is a graph Laplacian Δ_m defined on each of the graphs Γ_m,

$$(2.2.2) \qquad \Delta_m u(x) = \sum_{y \underset{m}{\sim} x} (u(y) - u(x)), \qquad x \in V_m \setminus V_0,$$

so, at least for $x \in V_* \setminus V_0$, we would like $\Delta_\mu u(x)$ to be a renormalized limit of $\Delta_m u(x)$. Such a formula would imply the local property of the operator Δ_μ, namely, that $\Delta_\mu u(x)$ is determined by the values of u in any neighborhood of x. This property is sometimes expressed more formally by the implication that $u \cdot v \equiv 0$ implies $v\Delta_\mu u \equiv 0$.

There is a rather simple way to derive such a pointwise formula from the weak formulation; it is the same idea used in the proof of Theorem 2.1.2, namely, to substitute $v = \psi_x^{(m)}$ into the definition. Note that $\mathcal{E}_m(u, \psi_x^{(m)}) = -r^{-m}\Delta_m u(x)$, since only edges containing x see any change in $\psi_x^{(m)}$. Also by Lemma 1.3.1, we know $\mathcal{E}_m(u, \psi_x^{(m)}) = \mathcal{E}(u, \psi_x^{(m)})$. Thus (2.1.4) says

$$(2.2.3) \qquad r^{-m}\Delta_m u(x) = \int_K f\psi_x^{(m)} d\mu.$$

Since f is assumed continuous and the support of $\psi_x^{(m)}$ is close to x,

$$\int_K f\psi_x^{(m)} d\mu \approx f(x) \int_K \psi_x^{(m)} d\mu.$$

The pointwise formula we are looking for then must be

$$(2.2.4) \qquad \Delta_\mu u(x) = \lim_{m \to \infty} r^{-m} \left(\int_K \psi_x^{(m)} d\mu \right)^{-1} \Delta_m u(x).$$

THEOREM 2.2.1 *Let $u \in dom\, \Delta_\mu$. Then the pointwise formula (2.2.4) holds with the limit uniform across $V_* \setminus V_0$. Conversely, suppose u is a continuous function and the right side of (2.2.4) converges uniformly to a continuous function on $V_* \setminus V_0$. Then $u \in dom\, \Delta_\mu$ and (2.2.4) holds.*

Proof: Suppose $u \in dom_\mu \Delta$ and $\Delta_\mu u = f$. Using (2.2.3) we find

$$(2.2.5) \qquad r^{-m} \left(\int_K \psi_x^{(m)} d\mu \right)^{-1} \Delta_m u(x) = \frac{\int_K f \psi_x^{(m)} d\mu}{\int_K \psi_x^{(m)} d\mu}$$

for any $x \in V_m \setminus V_0$. The uniform convergence of the right side of (2.2.5) to $f(x)$ is a routine consequence of the continuity of f.

Conversely, suppose the right side of (2.2.4) converges uniformly to a continuous function f. For any $v \in dom_0 \mathcal{E}$ we have

$$(2.2.6) \qquad \mathcal{E}_m(u, v) = r^{-m} \sum_{x \sim_m y} (u(y) - u(x))(v(y) - v(x)).$$

We collect all terms in the sum that contain the factor $v(x)$ for a fixed $x \in V_m \setminus V_0$ (since v vanishes on V_0, we can exclude boundary points). Such a factor will multiply $-(u(y) - u(x))$ for each y neighboring x. Thus

$$(2.2.7) \qquad \begin{aligned} \mathcal{E}_m(u, v) &= -r^{-m} \sum_{x \in V_m \setminus V_0} v(x) \left(\sum_{y \sim_m x} (u(y) - u(x)) \right) \\ &= - \sum_{x \in V_m \setminus V_0} v(x) r^{-m} \Delta_m u(x). \end{aligned}$$

(You might worry that we have thrown away half the terms in (2.2.6), namely the ones with factor $v(y)$, but in fact each edge in Γ_m appears twice in (2.2.7) and only once in (2.2.6).) So we can rewrite (2.2.2) as

$$(2.2.8) \qquad \mathcal{E}_m(u, v) = - \sum_{x \in V_m \setminus V_0} v(x) \left(\int \psi_x^{(m)} d\mu \right) r^{-m} \left(\int \psi_x^{(m)} d\mu \right)^{-1} \Delta_m u(x).$$

Define $f_m(x) = r^{-m} \left(\int \psi_x^{(m)} d\mu \right)^{-1} \Delta_m u(x)$ on $V_m \setminus V_0$. Then $f_m(x) \to f(x)$ uniformly by assumption, and (2.2.8) becomes

$$(2.2.9) \qquad \mathcal{E}_m(u, v) = - \int \left(\sum_{x \in V_m \setminus V_0} v(x) f_m(x) \psi_x^{(m)} \right) d\mu.$$

A routine argument (see exercises) shows that

$$(2.2.10) \qquad \sum_{x \in V_m \setminus V_0} v(x) f_m(x) \psi_x^{(m)}$$

converges uniformly to $v(x)f(x)$, so we may take the limit in (2.2.9) as $m \to \infty$ to obtain (2.1.4). $\qquad\square$

The converse statement is not true without the assumption that the limit is uniform. See the exercises for a counterexample.

Since the limit is uniform, we can easily "fill in" for values of x not in $V_* \setminus V_0$, namely

$$(2.2.11) \qquad \Delta_\mu u(x) = \lim_{m \to \infty} r^{-m} \left(\int \psi_{x_m}^{(m)} d\mu \right)^{-1} \Delta_m u(x_m),$$

where $\{x_m\}$ is any sequence $x_m \in V_m \setminus V_0$ with $\lim_{m \to \infty} x_m = x$. In particular, this is the only pointwise formula for $x \in V_0$.

In the case of the standard measure, it is not difficult to compute the factor $\left(\int \psi_x^{(m)} d\mu \right)^{-1}$ exactly, and the answer is independent of x. For the case $K = I$, $\psi_x^{(m)}$ is just a tent function, so $\int \psi_x^{(m)} dx$ is just the area of a triangle of height 1 and base $2 \cdot \frac{1}{2^m}$. So, $\left(\int \psi_x^{(m)} dx \right)^{-1} = 2^m$ and $r^{-m} = 2^m$, and so the pointwise formula is exactly (2.2.1). For the case $K = SG$, the function $\psi_x^{(m)}$ is supported in the two m-cells meeting at x. If $F_w K$ is one of these cells with vertices x, y, and z, then $\psi_x^{(m)} + \psi_y^{(m)} + \psi_z^{(m)}$ restricted to $F_w K$ is identically 1 (it is harmonic and takes on the value 1 at all three vertices). Thus

$$\int_{F_w K} (\psi_x^{(m)} + \psi_y^{(m)} + \psi_z^{(m)}) \, d\mu = \mu(F_w K) = \frac{1}{3^m}.$$

By symmetry all three summands have the same integral, so $\int_{F_w K} \psi_x^{(m)} d\mu = \frac{1}{3^{m+1}}$. Together with the contribution from the other m-cell we find $\int \psi_x^{(m)} d\mu = \frac{2}{3} \cdot \frac{1}{3^m}$, hence $\left(\int \psi_x^{(m)} d\mu \right)^{-1} = \frac{3}{2} \cdot 3^m$. Since $r^{-m} = (\frac{5}{3})^m$ in this case, we find

$$(2.2.12) \qquad \Delta u(x) = \frac{3}{2} \lim_{m \to \infty} 5^m \Delta_m u(x).$$

This particular pointwise formula will play a key role in Chapter 3.

Some people have wondered about the significance of the renormalization factor 5 in (2.2.12), and to a lesser extent the initial factor $\frac{3}{2}$. One could simply regard $\frac{3}{2}$ as a normalization constant (although it is explainable in terms of the choice of normalization constants in the energy and measure), and clearly it plays no significant role in the theory. Some references omit it altogether (this creates minor headaches in comparing formulas from different papers). But the factor of 5 is crucial. Suppose we try to replace it by a different factor, say R. If $R < 5$, then

$$\lim_{m \to \infty} R^m \Delta_m u(x) = 0 \quad \text{for all } u \in \text{dom}\, \Delta.$$

So we would end up with a theory in which virtually every function has Laplacian identically zero. Not a very interesting theory! On the other hand, if $R > 5$ then $\lim_{m \to \infty} R^m \Delta_m u(x)$ does not exist for most functions in dom Δ. Not a very interesting theory either!

It is also possible to compute the factors exactly in the case that μ is a self-similar measure. In this case the factor does depend on x. We leave the details to the exercises.

EXERCISES

2.2.1. Show that for $K = I$, the function $u(x) = |x - \frac{1}{3}|$ has the limit in (2.2.4) equal to 0 everywhere, but the limit is not uniform and u is not C^2.

2.2.2. For $K = I$, show that the function $u(x) = (x - \frac{1}{2})|x - \frac{1}{2}|$ has the limit in (2.2.4) converging uniformly to a discontinuous function, and u is not C^2.

2.2.3. Prove that

$$\sum_{x \in V_m \setminus V_0} v(x) f_m(x) \psi_x^{(m)}$$

converges uniformly to $v(x) f(x)$, where v, f, and f_m are as in the proof of Theorem 2.2.1 (in fact, this argument only requires v to be continuous).

2.2.4. Give a complete proof of (2.2.11).

2.2.5. Use the results of Exercise 1.2.5 to compute $\int \psi_x^{(m)} d\mu$ when μ is a self-similar measure on I.

2.2.6. Let μ be any self-similar measure on SG. Compute $\int_{SG} h d\mu$ for any harmonic functions in terms of the boundary values, by using the self-similar identity (1.2.19) and the harmonic extension algorithm (1.3.25) (see Exercise 1.3.10).

2.2.7. Use the results of Exercise 2.2.6 to compute $\int \psi_x^{(m)} d\mu$ when μ is a self-similar measure on SG.

2.2.8. Show that if u is the restriction to SG of a nonconstant linear function then u is not in dom Δ. Do the same if u is the restriction of a nonconstant C^2 function.

2.3 NORMAL DERIVATIVES

We began the chapter with an integration-by-parts formula (2.1.1) stripped of the boundary terms by the assumption that v vanishes on the boundary. If we drop this assumption, the full formula is

$$(2.3.1) \qquad \int_0^1 u''(x)v(x)dx = -\int_0^1 u'(x)v'(x)dx + u'(1)v(1) - u'(0)v(0).$$

To restore the symmetry between the boundary points we define the normal derivatives $\partial_n u(1) = u'(1)$ and $\partial_n u(0) = -u'(0)$ to measure the rate of change in a direction moving outside of I. Then we can rewrite (2.3.1) as

$$(2.3.2) \qquad \mathcal{E}(u, v) = -\int_I (\Delta u)v d\mu + \sum_{\partial I} v \partial_n u,$$

where μ is the standard measure. In this form it is just a special case of the Gauss-Green formula

$$(2.3.3) \qquad \int_\Omega \nabla u \cdot \nabla v d\mu = -\int_\Omega (\Delta u)v d\mu + \int_{\partial\Omega} v \partial_n u d\sigma$$

for an open set Ω in $\mathbb{R}^n$, with $d\sigma$ the surface measure on $\partial\Omega$. Here we are seeking both a Gauss–Green formula

$$(2.3.4) \qquad \mathcal{E}(u, v) = -\int_K (\Delta_\mu u)v \, d\mu + \sum_{V_0} v \partial_n u$$

and a definition of $\partial_n u$ at boundary points, so that (2.3.4) becomes (2.3.2) in the case $K = I$.

It will become apparent that the Gauss–Green formula and the definition (and existence) of normal derivatives go hand in hand. Nevertheless, the definition of the normal derivatives depends on the energy alone. Thus we have the paradoxical situation that $u \in \mathrm{dom}\,\Delta_\mu$ for any measure μ implies the existence of the same normal derivatives (perhaps the correct way to look at it is that $u \in \mathrm{dom}_{\mathcal{M}}\Delta$ is the condition that implies the existence of normal derivatives). In the case $K = I$, the standard Laplacian is a second derivative and the normal derivatives are first derivatives. In the case $K = SG$ you might want to think of $\partial_n u$ as a first derivative, but it is surely not the restriction to the boundary of any globally defined local operator. At any rate, it will be clear that the existence of normal derivatives is a weaker condition than the existence of a Laplacian. Also, having finite energy does not imply the existence of normal derivatives.

The key idea is to return to the proof of Theorem 2.2.1, but now do not require v to vanish at the boundary. Then we still have (2.2.6), but when we rearrange terms as in (2.2.7), we will also have terms with a factor $v(x)$ for $x \in V_0$, which we treat separately:

$$\mathcal{E}_m(u, v) = - \sum_{x \in V_m \setminus V_0} v(x) r^{-m} \Delta_m u(x)$$

$$(2.3.5)$$

$$+ \sum_{x \in V_0} v(x) r^{-m} \left(\sum_{y \underset{m}{\sim} x} (u(x) - u(y)) \right).$$

In a sense, this is a discrete analog of the Gauss–Green formula, with

$$(2.3.6) \qquad r^{-m} \left(\sum_{y \underset{m}{\sim} x} (u(x) - u(y)) \right)$$

playing the role of the normal derivative. Note that the number of summands in (2.3.6) is half the number in the definition of the discrete Laplacian, because boundary points have half the number of neighbors as junction points. Also, we have switched $u(y) - y(x)$ to $u(x) - u(y)$, absorbing the minus sign that stands in front of the first term on the right side of (2.3.5). We want to take the limit as $m \to \infty$ in (2.3.5), so we need to define the normal derivative as the limit of (2.3.6).

DEFINITION 2.3.1 Let $x \in V_0$ and u be a continuous function of K. We say $\partial_n u(x)$ exists if the limit of (2.3.6) exists, and

$$(2.3.7) \qquad \partial_n u(x) = \lim_{m \to \infty} r^{-m} \sum_{y \underset{m}{\sim} x} (u(x) - u(y)).$$

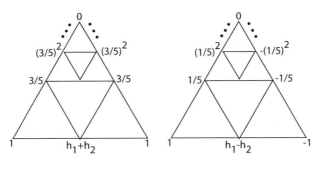

Figure 2.3.1

THEOREM 2.3.2 *Suppose $u \in \text{dom } \Delta_\mu$ for some measure μ. Then $\partial_n u(x)$ exists for all $x \in V_0$, and the Gauss–Green formula (2.3.4) holds for all $v \in \text{dom} \mathcal{E}$.*

Proof: First choose $v \in \text{dom } \mathcal{E}$ such that $v(q_i) = 1$ and $v(q_j) = 0$ for $j \neq i$. Then in the second term on the right side of (2.3.5), only one summand survives, namely $r^{-m} \sum_{y \widetilde{\sim} q_i} (u(q_i) - u(y))$. Now the left side converges to $\mathcal{E}(u, v)$, and the first term on the right side converges to $-\int v \Delta_\mu u d\mu$ as shown in the proof of Theorem 2.2.1. It follows that the limit (2.3.7) exists at $x = q_i$, so the normal derivative exists at every boundary point. Finally, for any $v \in \text{dom } \mathcal{E}$, we may take the limit in (2.3.5) to obtain (2.3.4). $\square$

For $K = I$ we recognize the limit (2.3.7) as defining an ordinary first derivative. For $K = SG$, we can write (2.3.7) explicitly as

$$(2.3.8) \qquad \partial_n u(q_i) = \lim_{m \to \infty} \left(\frac{5}{3}\right)^m (2u(q_i) - u(F_i^m q_{i+1}) - u(F_i^m q_{i-1})).$$

Note that the factor $(\frac{5}{3})^m$ in (2.3.8) is smaller than the factor 5^m in the pointwise formula (2.2.11) for the Laplacian. This is in keeping with the notion that the normal derivative is a "lower order" derivative than the Laplacian.

We want to compute the normal derivatives of harmonic functions on SG. For simplicity we do the computation at q_0. Recall the three eigenfunctions $h_0 + h_1 + h_2 \equiv 1$, $h_1 + h_2$ and $h_1 - h_2$ for the mapping $u \to u \circ F_0$ (see Figure 2.3.1).

For the constant function all the values of $(\frac{5}{3})^m (2u(q_0) - u(F_0^m q_1) - u(F_0^m q_2))$ are zero, and so the normal derivative is zero. The same is true for the skew-symmetric function $h_1 - h_2$, where $u(q_0) = 0$ and $u(F_0^m q_2) = -u(F_0^m q_1)$. So the only interesting computation is for $u = h_1 + h_2$. Here $u(q_0) = 0$ and $u(F_0^m q_1) = u(F_0^m q_2) = (\frac{3}{5})^m$ because the eigenvalue is $\frac{3}{5}$, so $(\frac{5}{3})^m (2u(q_0) - u(F_0^m q_1) - u(F_0^m q_2)) = -2(\frac{5}{3})^m (\frac{3}{5})^m = -2$, and so $\partial_n u(q_0) = -2$. Notice that in all three cases we do not have to take the limit to compute the normal derivative because the sequence is constant. This is true for all harmonic functions by linearity:

$$(2.3.9) \qquad \partial_n h(q_0) = 2h(q_0) - h(q_1) - h(q_2).$$

Wow, that's easy!

We can also localize the definition of normal derivative. Let $F_w K$ be any cell, and let $x = F_w q_i$ be a boundary point of the cell. Then we may define $\partial_n u(x)$

with respect to the cell $F_w K$ by the same formula (2.3.7), with the understanding that neighbors y on the right side are restricted to lie in $F_w K$ (so there are always two neighbors in the case of SG). Another way of saying this is

$$(2.3.10) \qquad \partial_n u(F_w q_i) = r^{-|w|} \partial_n (u \circ F_w)(q_i).$$

This notation suggests that we are working in the cell $F_w K$. If x is a junction point then $x = F_{w'} q_{i'}$ for another word w' with $|w'| = |w|$, and so there is another local normal derivative $\partial_n u(F_{w'} q_{i'})$ at x with respect to the cell $F_{w'} q_{i'}$. These normal derivatives may be unrelated, but in the interesting cases they sum to zero, so they just differ by a sign.

THEOREM 2.3.3 *Suppose $u \in \text{dom} \, \Delta_\mu$. Then at each junction point $x = F_w q_i = F_{w'} q_{i'}$, the local normal derivatives exist and*

$$(2.3.11) \qquad \partial_n u(F_w q_i) + \partial_n u(F_{w'} q_{i'}) = 0.$$

This is called the matching condition for normal derivatives at x.

Proof: The existence follows from (2.3.10). Now

$$(2.3.12) \qquad \partial_n u(F_w q_i) + \partial_n u(F_{w'} q_{i'}) = \lim_{m \to \infty} r^{-m} \sum_{y \underset{m}{\sim} x} (u(x) - u(y))$$

since the neighbors y of x lie in either $F_w K$ or $F_{w'} K$. Since $u \in \text{dom} \, \Delta_\mu$ we know that

$$\lim_{m \to \infty} r^{-m} \left(\int \psi_x^{(m)} d\mu \right)^{-1} \Delta_m u(x)$$

exists, and since $\int \psi_x^{(m)} d\mu \to 0$ as $m \to \infty$ (this follows from the continuity property (1.2.3) of the measure) we see that the limit on the right side of (2.3.12) is zero, proving (2.3.11). $\qquad \qquad \square$

It might seem that the local normal derivative at all junction points in SG comes very close to a derivative definable on all of SG, at least if we consider $|\partial_n u(x)|$ to avoid the ambiguity in sign. But this is misleading, as the function $|\partial_n u(x)|$ behaves in a very discontinuous manner as x varies, and so cannot be extended to points not in V_*.

Similarly, we can localize the Gauss–Green formula to any cell $F_w K$,

$$(2.3.13) \qquad \mathcal{E}_{F_w K}(u, v) = -\int_{F_w K} (\Delta_\mu u) v \, d\mu + \sum_{\partial F_w K} v \partial_n u,$$

or to any finite union of cells A by addition,

$$(2.3.14) \qquad \mathcal{E}_A(u, v) = -\int_A (\Delta_\mu u) v \, d\mu + \sum_{\partial A} v \partial_n u,$$

for the appropriate definition of ∂A (note that if two cells in A intersect at a junction point x, then x will not be in ∂A, but the contributions at x of $v \partial_n u$ will cancel by the matching condition (2.3.11)).

EXERCISES

2.3.1. Show that the function $u(x) = x^{2/3}$ on I belongs to dom $\mathcal{E}$, but $\partial_n u(0)$ does not exist.

2.3.2. Show that $\sum_i \partial_n h(q_i) = 0$ for any harmonic function.

2.3.3. Compute the local normal derivatives $\partial_n h_0(x)$ for all $x \in V_2$. Note the jumpy behavior of $|\partial_n h_0(x)|$ as x varies.

2.3.4. Suppose we minimize energy on SG over all functions u with $u(q_1)$ and $u(q_2)$ specified. Show that we obtain a harmonic function satisfying $\partial_n u(q_0) = 0$.

2.3.5. Let h be a harmonic function on SG with $\partial_n h(q_0) \neq 0$. Show that

$$\lim_{m \to \infty} \left(\frac{5}{3}\right)^m (h(q_0) - h(F_0^m q_1)) \quad \text{and} \quad \lim_{m \to \infty} \left(\frac{5}{3}\right)^m (h(q_0) - h(F_0^m q_2))$$

both exist and are different from 0. Conclude that the restriction of h to the line segments joining q_1 to q_0 (or q_2 to q_0) is not differentiable (in the usual sense) at the point q_0.

2.3.6. Let h be a harmonic function on SG with $\partial_n u(q_0) = 0$. Show that $\lim_{m \to \infty} 5^m (h(q_0) - h(F_0^m q_1))$ and $\lim_{m \to \infty} 5^m (h(q_0) - h(F_0^m q_2))$ exist. Show that the restriction of h to the line segments joining q_1 to q_0 (or q_2 to q_0) is differentiable at the point q_0 with derivative equal to 0.

2.3.7. If $\partial_n u(q_0)$ exists on SG, show that $\partial_n (u \circ F_0^m)(q_0)$ exists and $\partial_n (u \circ F_0^m)(q_0) = (\frac{3}{5})^m \partial_n u(q_0)$.

2.3.8. Let u be a function on SG that is skew-symmetric with respect to the reflection R_0 preserving q_0. Show that $\partial_n u(q_0) = 0$. Conclude that for a general function u, the existence and value of $\partial_n u(q_0)$ depends only on the symmetric part $\frac{1}{2}(u(x) + u(R_0 x))$ of u.

2.3.9.* Show that the conclusions of Theorem 2.3.2 are valid if we only assume $u \in \text{dom}_{L^2} \Delta_\mu$, or even just $u \in \text{dom}_{\mathcal{M}} \Delta$.

2.3.10. Suppose u is continuous on SG and $u \circ F_i$ is harmonic for $i = 0, 1, 2$. Show that $u \in \text{dom } \Delta_\mu$ if and only if u is harmonic.

2.4 GAUSS–GREEN FORMULA

We have already seen one version of the Gauss–Green formula in Section 2.3, in both the global form (2.3.4) and the local form (2.3.14). There is another, more symmetric version that does not involve energy at all.

Theorem 2.4.1 *Suppose u and v are in dom Δ_μ. Then*

$$(2.4.1) \qquad \int_K u \Delta_\mu v \, d\mu - \int_K v \Delta_\mu u \, d\mu = \sum_{V_0} (u \partial_n v - v \partial_n u).$$

Locally, if A denotes any finite union of cells with boundary ∂A,

$$(2.4.2) \qquad \int_A u \Delta_\mu v \, d\mu - \int_A v \Delta_\mu u \, d\mu = \sum_{\partial A} (u \partial_n v - v \partial_n u).$$

Proof: Take (2.3.4) for u and v in both orders, and subtract. $\qquad\square$

This version of the Gauss–Green formula is a versatile tool for understanding the Laplacian. In typical applications we will make specific choices of the function v. For example, if we choose $v \equiv 1$ then we obtain

$$(2.4.3) \qquad \int_K \Delta_\mu u\, d\mu = \sum_{V_0} \partial_n u$$

or the local version

$$(2.4.4) \qquad \int_A \Delta_\mu u\, d\mu = \sum_{\partial A} \partial_n u.$$

The special case when u is harmonic, so the left side is zero, was given in Exercise 2.3.2. One consequence is that if $u \in$ dom Δ_μ and $\partial_n u(x) = 0$ for all $x \in V_*$, then u is constant. Indeed, in that case $\int_{F_w K} \Delta_\mu u\, d\mu = 0$ for all cells $F_w K$, which implies $\Delta_\mu u = 0$ so u is harmonic. But we know how to compute normal derivatives of harmonic functions at boundary points, and we don't get all zeros unless the function is constant. Similarly, if $\partial_n u(x) = 0$ for all $x \in A \cap V_*$, for a connected set A, then u is constant on A.

The next application gives a quantitative version of the pointwise formula for the Laplacian.

THEOREM 2.4.2 *If $u \in$ dom Δ_μ, then*

$$(2.4.5) \qquad r^{-m}\left(\int \psi_x^{(m)} d\mu\right)^{-1} \Delta_m u(x) = \frac{\int \psi_x^{(m)} \Delta_\mu u\, d\mu}{\int \psi_x^{(m)} d\mu}$$

for $x \in V_m \setminus V_0$.

Proof: We give the argument for $K = SG$. Suppose $x = F_w q_i = F_{w'} q_{i'}$ for $|w| = |w'| = m$. We will use (2.4.2) where $A = F_w K$ and v is the harmonic function with $v \circ F_w = h_i$, so $v(F_w q_j) = \delta_{ij}$, $\partial_n v(F_w q_i) = 2r^{-m}$ and $\partial_n v(F_w q_j) = -r^{-m}$ for $j \neq i$. Thus

$$(2.4.6) \qquad \begin{aligned} \int_{F_w K} \psi_x^{(m)} \Delta_\mu u\, d\mu &= \partial_n u(F_w q_i) - 2r^{-m} u(F_w q_i) \\ &\quad + r^{-m} u(F_w q_{i+1}) + r^{-m} u(F_w q_{i-1}). \end{aligned}$$

We do a similar computation on $F_{w'} K$ and add the two. Note that $\partial_n u(F_w q_i) + \partial_n u(F_{w'} q_{i'}) = 0$ by Theorem 2.3.3. Thus we obtain

$$(2.4.7) \qquad \int \psi_x^{(m)} \Delta_\mu u\, d\mu = r^{-m} \Delta_m u(x),$$

which is the same as (2.4.5). $\qquad\square$

The right side of (2.4.5) is just an average value of $\Delta_\mu u$ over a small set supp $\psi_x^{(m)}$ containing x, so we recover the pointwise formula in the limit. However, (2.4.5) allows us to estimate the error, and if we assume some smoothness for $\Delta_\mu u$

(such as a Hölder condition) we obtain a rate of convergence for the pointwise formula.

The same reasoning will give a rate of convergence in the definition of the normal derivative. For example, if $x = q_0$ then we use (2.4.6) on $F_0^m K$ to obtain

$$(2.4.8) \quad r^{-m}(2u(q_0) - u(F_0^m q_1) - u(F_0^m q_2)) = \partial_n u(q_0) - \int_{F_0^m K} \psi_{q_0}^{(m)} \Delta_\mu u d\mu.$$

For the standard Laplacian this gives a rate $O\left(\frac{1}{3^m}\right)$ for the convergence of the left side of (2.4.8) to $\partial_n u(q_0)$. We could also express this as

$$(2.4.9) \qquad 2u(q_0) - u(F_0^m q_1) - u(F_0^m q_2) = \left(\frac{3}{5}\right)^m \partial_n u(q_0) + O\left(\frac{1}{5^m}\right),$$

or more precisely

$$(2.4.10) \qquad \begin{aligned} 2u(q_0) - u(F_0^m q_1) - u(F_0^m q_2) &= \left(\frac{3}{5}\right)^m \partial_n u(q_0) \\ &\quad - \frac{1}{3}\left(\frac{1}{5^m}\right)\Delta u(q_0) + o\left(\frac{1}{5^m}\right) \end{aligned}$$

for $u \in \mathrm{dom}\,\Delta$.

The symmetric form of the Gauss–Green formula allows us to conclude that Δ_μ is a symmetric operator with respect to the inner product

$$(2.4.11) \qquad \langle u, v \rangle = \int_K uv d\mu$$

provided we impose suitable boundary conditions. For example, the *Dirichlet domain* of Δ_μ is the codimension 3 subspace of dom Δ_μ of functions vanishing on V_0. Note that if both u and v are in the Dirichlet domain then the boundary terms in (2.4.1) vanish, so we obtain

$$(2.4.12) \qquad \langle \Delta_\mu u, v \rangle = \langle u, \Delta_\mu v \rangle.$$

Similarly, the *Neumann domain* is defined by the vanishing of the normal derivatives on the boundary, and (2.4.12) continues to hold if both u and v are in the Neumann domain. Other boundary conditions also produce symmetry, such as putting Dirichlet conditions on some boundary points and Neumann conditions on the others. Later we will show that the Laplacian is essentially self-adjoint on these domains, and the domain of the closure is $\mathrm{dom}_{L^2}\,\Delta_\mu$ with the given boundary conditions.

EXERCISES

2.4.1. If $u \in \mathrm{dom}\,\Delta_\mu$ on SG and $\Delta_\mu u$ is Hölder continuous of order α in the Euclidean metric, show that $r^{-m}\left(\int \psi_x^{(m)} d\mu\right)^{-1} \Delta_m u(x) = \Delta_\mu(x) + O\left(\frac{1}{2^{m\alpha}}\right)$.

2.4.2. If $\Delta_\mu u \geq 0$ on K, show that $\Delta_m u(x) \geq 0$ for all $x \in V_m \setminus V_0$.

2.4.3. If $\Delta_\mu u \geq 0$ on SG, show that $r^{-m}(2u(q_0) - u(F_0^m q_1) - u(F_0^m q_2)) \leq \partial_n u(q_0)$ for all m.

2.4.4. For $u \in \text{dom } \Delta$, show that

$$\Delta u(q_0) = \lim_{m \to \infty} \left(3^{m+1} \partial_n u(q_0) - 3 \cdot 5^m (2u(q_0) - u(F_0^m q_1) - u(F_0^m q_2))\right).$$

2.4.5. Show that (2.4.12) holds if u and v both belong to the subspace of dom Δ_μ satisfying the boundary conditions $a_i u(q_i) + b_i \partial_n u(q_i) = 0$ at each q_i, where $(a_i, b_i) \neq (0, 0)$.

2.4.6. If $\Delta_\mu u \geq 0$ but not identically zero, show that $\partial_n u(q_i) > 0$ for at least one boundary point.

2.4.7.* Show that (2.4.1) holds if u and v belong to $\text{dom}_{L^2} \Delta_\mu$.

2.5 GLUING

As every calculus student should learn, functions may be defined in pieces. For example, suppose f_0 is a continuous function on $[0, \frac{1}{2}]$ and f_1 is a continuous function on $[\frac{1}{2}, 1]$. Then we can try to glue them together to create a continuous function on I,

$$(2.5.1) \qquad\qquad f(x) = \begin{cases} f_0(x), & x \in [0, \frac{1}{2}], \\ f_1(x), & x \in [\frac{1}{2}, 1]. \end{cases}$$

Of course there is an ambiguity in the definition of f at the junction point $\frac{1}{2}$. In a typical calculus text this would be carefully evaded by rewriting the definition to arbitrarily exclude $\frac{1}{2}$ from one or the other interval. But if we want f to be continuous we must impose the *matching condition* $f_0(\frac{1}{2}) = f_1(\frac{1}{2})$, and this also resolves the ambiguity. If f_0 and f_1 are C^1 and we want f to be C^1, we need to impose a second matching condition $f_0'(\frac{1}{2}) = f_1'(\frac{1}{2})$. Note that we may also write this as

$$(2.5.2) \qquad\qquad \partial_n f(F_0(1)) + \partial_n f(F_1(0)) = 0,$$

since $\partial_n f(F_1(0)) = -f_1'(\frac{1}{2})$. If f_1 and f_2 are C^2 and we want f to be C^2, then we must add a third matching condition $f_0''(\frac{1}{2}) = f_1''(\frac{1}{2})$. These conditions are clearly necessary and sufficient.

In this section we want to answer similar questions for functions defined in pieces on SG. We will discuss this for the decomposition

$$(2.5.3) \qquad\qquad K = \bigcup_i F_i K$$

into cells of level 1, but the same results are valid for any decomposition

$$K = \bigcup_{w \in \mathcal{P}} F_w K$$

into cells for any partition $\mathcal{P}$, as discussed in Exercise 1.2.2. So we assume that we are given continuous functions f_i defined on $F_i K$, and we glue them to form

$$(2.5.4) \qquad\qquad f(x) = \{f_i(x) \text{ if } x \in F_i K\}.$$

Clearly f is well defined and continuous if and only if the matching conditions

(2.5.5) $$f_i(F_i q_j) = f_j(F_j q_i) \quad \text{for } i \neq j$$

hold. These conditions also suffice to guarantee that $f \in \text{dom } \mathcal{E}$ if the functions f_i have finite energy. What does this mean? We can say either that $f_i \circ F_i$ as a function on K has finite energy or that the energy

(2.5.6) $$\mathcal{E}_{F_i K}(f_i) = \lim_{m \to \infty} r^{-m} \sum_{\substack{x \sim y \\ m \\ \text{in } F_i K}} (f_i(x) - f_i(y))^2$$

is finite. Because the sum defining $\mathcal{E}_m(f)$ splits exactly into contributions from the cells $F_i K$, all these statements are straightforward.

Next we look at what happens for dom Δ_μ. Suppose we have continuous functions u_i and f_i defined on $F_i K$ such that $\Delta_\mu u_i = f_i$ on $F_i K$. If we glue them together to create functions u and f, do we have $\Delta_\mu u = f$? Before tackling this question we need to clarify the meaning of the statement $\Delta_\mu u_i = f_i$ on $F_i K$. We could simply localize the weak formulation:

(2.5.7) $$\mathcal{E}_{F_i K}(u_i, v) = -\int_{F_i K} f_i v \, d\mu \quad \text{for all } v \in \text{dom } \mathcal{E} \text{ vanishing on } \partial F_i K.$$

We could also restate this in global terms as

$$\mathcal{E}(u_i, v) = -\int_K f_i v \, d\mu \quad \text{for all } v \in \text{dom}_0 \mathcal{E} \text{ with support in } F_i K.$$

The weak formulation is then equivalent to the pointwise formula (2.2.4) holding uniformly on $V_* \cap F_i K \setminus \partial F_i K$. For the standard Laplacian we can also show that this is equivalent to $u \circ F_i \in \text{dom } \Delta$ with $\Delta(u \circ F_i) = \frac{1}{5} f_i \circ F_i$.

In order for u and f to be continuous, they both must satisfy the matching conditions (2.5.5). But this alone is not sufficient, because if $u \in \text{dom } \Delta_\mu$, then by Theorem 2.3.3 the normal derivatives at junction points must satisfy matching conditions. At the point $F_i q_j = F_j q_i$, this condition in terms of the functions u_i may be written

(2.5.8) $$\partial_n u_i(F_i q_j) + \partial_n u_j(F_j q_i) = 0 \quad \text{for } i \neq j.$$

This is clearly the analog of (2.5.2).

THEOREM 2.5.1 *Let u and f be defined by gluing pieces $\{u_i\}$ and $\{f_i\}$, with $\Delta_\mu u_i = f_i$ on $F_i K$. Then $u \in \text{dom } \Delta_\mu$ with $\Delta_\mu u = f$ if and only if (2.5.5) holds for $\{u_i\}$ and $\{f_i\}$ (so u and f are continuous), and the matching conditions on normal derivatives (2.5.8) hold.*

Proof: We have already seen that the conditions are necessary, so assume that they hold. We will show that the weak formulation of $\Delta_\mu u = f$ is valid. Choose $v \in \text{dom}_0 \mathcal{E}$. We can apply the local Gauss–Green formula (2.3.13) on each cell $F_i K$ to obtain

(2.5.9) $$\mathcal{E}_{F_i K}(u_i, v) = -\int_{F_i K} f_i v \, d\mu + \sum_{\partial F_i K} v \partial_n u_i,$$

and then sum over i. The claim is that the boundary terms vanish in the sum. Indeed, this is true at points in V_0 because v vanishes there by assumption. At junction points $F_i q_j = F_j q_i$ there are two terms with the same v values, and so the sum vanishes by the matching condition (2.5.8). Thus the sum of (2.5.9) yields $\mathcal{E}(u, v) = -\int f v d\mu$ as required. $\qquad\qquad\qquad\square$

One could imagine trying to prove this theorem using the pointwise formula, but the result would be disappointing. Indeed, we already know that the pointwise formula holds uniformly for all points in $V_* \setminus V_0$ except the three junction points. Moreover, the matching condition at a junction point says

$$(2.5.10) \qquad\qquad \lim_{m \to \infty} \left(\frac{5}{3}\right)^m \Delta_m u(x) = 0.$$

This condition is clearly necessary for the pointwise formula, but it still falls short of the mark.

If we are willing to accept the weaker conclusion $u \in \mathrm{dom}_{L^2}\Delta_\mu$, then we don't need to assume the matching condition (2.5.5) for $\{f_i\}$. Then f is not necessarily continuous. (The correct statement would be to assume only that f_i belong to $L^2(F_i K, d\mu)$.) We will actually encounter this situation when we discuss biharmonic splines, where Δu has jump discontinuities at junction points.

Gluing is an important technical tool, but we need to first find a larger supply of functions in dom Δ_μ before we can use it effectively. It is the best available substitute for arguments in analysis on manifolds that use partitions of unity.

EXERCISES

2.5.1. Prove the equivalence of the weak formulation and the pointwise formula for the statement $\Delta_\mu u_i = f_i$ on $F_i K$.

2.5.2. For the standard Laplacian, show that $\Delta u_i = f_i$ on $F_i K$ is equivalent to $\Delta(u_i \circ F_i) = \frac{1}{5} f_i \circ F_i$.

2.5.3.* Suppose u_i are harmonic and satisfy (2.5.5), but not (2.5.8). Show that $u \in \mathrm{dom}_{\mathcal{M}}\Delta$ and $\Delta u = \nu$ for a discrete measure ν supported on the junction points.

2.5.4.* Suppose $u_i \in \mathrm{dom}_{L^2}\Delta_\mu$ and $\Delta_\mu u_i = f_i$ on $F_i K$ for $f_i \in L^2(F_i K, d\mu)$. Show that $u \in \mathrm{dom}_{L^2}\Delta_\mu$ and $\Delta_\mu = f$ if and only if u satisfies (2.5.2) and (2.5.8).

2.6 GREEN'S FUNCTION

The differential equation $-u'' = f$ on I, subject to Dirichlet boundary conditions $u(0) = u(1) = 0$, can be solved explicitly by

$$(2.6.1) \qquad\qquad u(x) = \int_0^1 G(x, y) f(y) dy,$$

for the Green's function

$$(2.6.2) \qquad G(x, y) = \begin{cases} x(1-y), & x \le y, \\ y(1-x), & y \le x. \end{cases}$$

There are various tricks to derive these formulas from scratch, but it is easy enough to check that (2.6.1) solves the problem and then show that the solution is unique. We will give another formula for $G(x, y)$ that exploits the self-similar structure of I and generalizes to SG. Note that (2.6.1) says that the operator $-\frac{d^2}{dx^2}$ mapping $C_D^2(I)$ (the subscript D indicates the boundary conditions $u(0) = u(1) = 0$) to $C(I)$ is invertible, and the inverse is the integral operator (2.6.1). Note that the explicit formula for the Green's function yields some surprises: It is positive, and it is continuous. The positivity is due to (and explains) the choice of the minus sign in the differential operator, and this is a general characteristic of Laplacians. The continuity is exceptional and is one of the manifestations of the fact that "points have positive capacity on I"; it is not true for Laplacians on higher dimensional manifolds. Another surprising observation is the symmetry, $G(x, y) = G(y, x)$. This can be explained by the symmetry of the differential operator on the space C_D^2.

To understand the Green's function we will break the symmetry and fix a value of y. What kind of a function (of x) is $G(\cdot, y)$? Clearly it is a piecewise linear function with a singularity at $x = y$, continuous there but not C^1. It has left and right derivatives, and the jump is

$$\frac{d}{dx}G(y^+, y) - \frac{d}{dx}G(y^-, y) = -y - (1-y) = -1.$$

At least formally this allows us to say that

$$(2.6.3) \qquad -\frac{d^2}{dx^2}G(x, y) = \delta(x - y),$$

and this explains why

$$-\frac{d^2}{dx^2}\int_0^1 G(x, y)f(y)dy = \int_0^1 \delta(x - y)f(y)dy = f(x).$$

Also, $G(x, y)$ clearly satisfies the boundary conditions $G(0, y) = G(1, y) = 0$, and so the integral (2.6.1) also satisfies the Dirichlet boundary conditions.

This explanation does not distinguish between generic points and dyadic rationals (points in V_*), so it will not generalize to SG. Instead, we want to work with the functions $\psi_x^{(m)}$ for $x \in V_m \setminus V_0$. These are piecewise linear "tent functions" with $\psi_x^{(m)}(x) = 1$, $\psi_x^{(m)}(x \pm \frac{1}{2^m}) = 0$, and $\psi_x^{(m)}$ vanishes outside $[x - \frac{1}{2^m}, x + \frac{1}{2^m}]$ (recall that $x = \frac{k}{2^m}$ for $1 \le k \le 2^m - 1$). If $y = \frac{1}{2}$ then $G(x, \frac{1}{2}) = \frac{1}{4}\psi_{1/2}^{(1)}(x)$, and we may write this as

$$G\left(x, \tfrac{1}{2}\right) = \tfrac{1}{4}\psi_{1/2}^{(1)}(x)\psi_{1/2}^{(1)}\left(\tfrac{1}{2}\right).$$

In general

$$(2.6.4) \qquad \frac{1}{4}\psi_{1/2}^{(1)}(x)\psi_{1/2}^{(1)}(y) = \left(\begin{cases} x & x \le \frac{1}{2} \\ 1-x & x \ge \frac{1}{2} \end{cases}\right)\left(\begin{cases} y & y \le \frac{1}{2} \\ 1-y & y \ge \frac{1}{2} \end{cases}\right).$$

Figure 2.6.1

Note that if $y > \frac{1}{2}$ this agrees with $G(x, y)$ for $x \le \frac{1}{2}$, while if $y < \frac{1}{2}$ it agrees with $G(x, y)$ for $x \ge \frac{1}{2}$. So the difference $G(x, y) - \frac{1}{4}\psi_{1/2}^{(1)}(x)\psi_{1/2}^{(1)}(y)$ is supported in either $F_0 I$ (if $y \in F_0 I$) or $F_1 I$ (if $y \in F_1 I$), and it still has a tent shape (see Figure 2.6.1).

Suppose $y < \frac{1}{2}$. Then

$$G(x, y) - \frac{1}{4}\psi_{1/2}^{(1)}(x)\psi_{1/2}^{(1)}(y) = \begin{cases} x(1-2y), & x \le y, \\ y(1-2x), & y \le x \le \frac{1}{2}, \\ 0, & x \ge \frac{1}{2}. \end{cases}$$

In other words, on $F_0 I$ this is exactly $\frac{1}{2}G(2x, 2y) = \frac{1}{2}G(F_0^{-1}x, F_0^{-1}y)$. Similarly, if $y > \frac{1}{2}$ then the difference on $F_1 I$ is $\frac{1}{2}G(F_1^{-1}x, F_1^{-1}y)$. Now if we are willing to make the convention that $G(x, y) = 0$ for x or y outside I, then we can write this concisely as

$$(2.6.5) \quad G(x, y) = \frac{1}{4}\psi_{1/2}^{(1)}(x)\psi_{1/2}^{(1)}(y) + \frac{1}{2}G(F_0^{-1}x, F_0^{-1}y) + \frac{1}{2}G(F_1^{-1}x, F_1^{-1}y).$$

The second term on the right is zero unless x and y both belong to $F_0 I$, and the third is zero unless x and y both belong to $F_1 I$. At most one is nonzero, and both are zero if x and y belong to different 1-cells.

We are now in a position to iterate (2.6.5). For simplicity of notation define

$$(2.6.6) \qquad \Psi(x, y) = \frac{1}{4}\psi_{1/2}^{(1)}(x)\psi_{1/2}^{(1)}(y).$$

Then

$$G(x, y) = \Psi(x, y) + \frac{1}{2}(G(F_0^{-1}x, F_0^{-1}y) + G(F_1^{-1}x, F_1^{-1}y))$$

$$= \Psi(x, y) + \frac{1}{2}(\Psi(F_0^{-1}x, F_0^{-1}y) + \Psi(F_1^{-1}x, F_1^{-1}y))$$

$$+ \frac{1}{4}(G(F_0^{-1}F_0^{-1}x, F_0^{-1}F_0^{-1}y) + \cdots),$$

and this leads to the infinite series

$$(2.6.7) \qquad G(x, y) = \sum_{m=0}^{\infty} \sum_{|w|=m} \left(\frac{1}{2}\right)^m \Psi(F_w^{-1}x, F_w^{-1}y).$$

The understanding is that each term $\Psi(F_w^{-1}x, F_w^{-1}y)$ is zero unless x and y both belong to the m-cell $F_w I$, so there is at most one nonzero term for each m. But also, if $x \ne y$, there are only a finite number of nonzero terms, since once $|x - y| > \frac{1}{2^m}$

they can't both belong to a single m-cell. Also, if $y \in V_m$, then the entire sum terminates at that value of m. Also, since $\psi_{1/2}^{(1)}(F_w^{-1}x) = \psi_z^{(m+1)}(x)$ for $z = F_w\frac{1}{2}$ and the points z obtained are exactly the points in $V_{m+1} \setminus V_m$, we may also write (2.6.7) as

$$(2.6.8) \qquad G(x, y) = \sum_{m=0}^{\infty} \sum_{z \in V_{m+1} \setminus V_m} \frac{1}{4} \left(\frac{1}{2}\right)^m \psi_z^{(m+1)}(x) \psi_z^{(m+1)}(y).$$

Now we would like to reverse the reasoning. In other words, we want to start with (2.6.8) as the definition of the Green's function and deduce (2.6.3) directly from it, without passing to (2.6.2). How does this work? The basic computation is that

$$
-\frac{d^2}{dx} \psi_z^{(m+1)}(x) = 4 \cdot 2^m \left(\delta(x - z) - \frac{1}{2}\delta\left(x - z - \frac{1}{2^{m+1}}\right) \right.
$$

$$(2.6.9)$$

$$
\left. - \frac{1}{2}\delta\left(x - z + \frac{1}{2^{m+1}}\right) \right).
$$

If one of the neighbors of z happens to be 0 or 1, the term $\delta(x)$ or $\delta(x - 1)$ does not really contribute because in the weak formulation we always integrate against a function v that vanishes on the boundary. If we take the sum on m in (2.6.8) up to $m = M$ and apply $-\frac{d^2}{dx^2}$, we obtain

$$
\sum_{m=0}^{M} \sum_{z \in V_{m+1} \setminus V_m} \left(\delta(x - z) - \frac{1}{2}\delta\left(x - z - \frac{1}{2^{m+1}}\right) \right.
$$

$$(2.6.10)$$

$$
\left. - \frac{1}{2}\delta\left(x - z + \frac{1}{2^{m+1}}\right) \right) \psi_z^{(m+1)}(y).
$$

This looks like a mess, but it actually simplifies if we rearrange the sum to group the terms containing $\delta(x - z)$ for fixed z. For example, when $M = 1$, we just have

$$
\delta\left(x - \frac{1}{2}\right) \left(\psi_{1/2}^{(1)}(y) - \frac{1}{2}\psi_{1/4}^{(2)}(y) - \frac{1}{2}\psi_{3/4}^{(2)}(y) \right)
$$

$$
+ \delta\left(x - \frac{1}{4}\right) \psi_{1/4}^{(2)}(y) + \delta\left(x - \frac{3}{4}\right) \psi_{3/4}^{(2)}(y),
$$

and moreover

$$
\psi_{1/2}^{(1)}(y) - \frac{1}{2}\psi_{1/4}^{(2)}(y) - \frac{1}{2}\psi_{3/4}^{(2)}(y) = \psi_{1/2}^{(2)}(y),
$$

because it is a piecewise linear function vanishing at 0, $\frac{1}{4}$, $\frac{3}{4}$, and 1 and taking the value of 1 at $\frac{1}{2}$. So (2.6.10) for $M = 1$ becomes

$$
\delta\left(x - \frac{1}{4}\right) \psi_{1/4}^{(2)}(y) + \delta\left(x - \frac{1}{2}\right) \psi_{1/2}^{(2)}(y) + \delta\left(x - \frac{3}{4}\right) \psi_{3/4}^{(2)}(y).
$$

By iterating the same reasoning we find

$$
\begin{aligned}
(2.6.11) \qquad &-\frac{d^2}{dx^2}\left(\sum_{m=0}^{M}\sum_{z\in V_{m+1}\backslash V_m}\frac{1}{4}\left(\frac{1}{2}\right)^m\psi_z^{(m+1)}(x)\psi_z^{(m+1)}(y)\right)\\
&= \sum_{z\in V_{M+1}\backslash V_0}\delta(x-z)\psi_z^{(M+1)}(y).
\end{aligned}
$$

What a beautiful formula! Note that if $y \in V_{M+1} \setminus V_0$ then there is only one nonzero term on the right side, namely $\delta(x-y)$. If not, say $z_0 < y < z_1$ with consecutive points $z_0, z_1 \in V_{M+1}$. Then there are only two nonzero terms,

$$\delta(x-z_0)\psi_{z_0}^{(M+1)}(y)+\delta(x-z_1)\psi_{z_1}^{(M+1)}(y).$$

Moreover $\psi_{z_0}^{(M+1)}(y)+\psi_{z_1}^{(M+1)}(y)=1$ on $[z_0,z_1]$, because it is a piecewise linear function equal to 1 at z_0 and z_1. So it is clear that the right side of (2.6.11) converges to $\delta(x-y)$ as $M\to\infty$.

Our goal is to replicate this wild construction on SG. How can we do this? Clearly we can't start at the beginning with an analog of (2.6.2), as we have no clue as to what it might be. Instead we will start at the end, with (2.6.11). Indeed, the right side of (2.6.11) makes perfect sense on SG and will lead to the same limit. We will need something a bit more elaborate on the left side, however, something like

$$
(2.6.12) \qquad \sum_{m=0}^{M}\sum_{z,z'\in V_{m+1}\backslash V_m} g(z,z')\psi_z^{(m+1)}(x)\psi_{z'}^{(m+1)}(y)
$$

for appropriately chosen coefficients $g(z,z')$. We will not succeed if we make $g(z,z')$ completely diagonal (vanishing unless $z=z'$), but it suffices to make it almost diagonal, with $g(z,z')=0$ unless z and z' belong to the same m-cell. We also expect a self-similar structure for the coefficients.

We begin with the case $m=0$. Then there are three points in $V_1 \setminus V_0$, and we write

$$
(2.6.13) \qquad \Psi(x,y)=\sum_{z,z'\in V_1\backslash V_0} g(z,z')\psi_z^{(1)}(x)\psi_{z'}^{(1)}(y).
$$

By symmetry we expect that

$$
(2.6.14) \qquad g(z,z)=a \quad\text{and}\quad g(z,z')=b \quad\text{if } z\neq z'
$$

for numbers a and b to be determined. By analogy with (2.6.7) we expect that

$$
(2.6.15) \qquad G(x,y)=\sum_{m=0}^{\infty}\sum_{|w|=m} r^m\Psi(F_w^{-1}x, F_w^{-1}y).
$$

This is the limit of expressions (2.6.12) with

$$
(2.6.16)
$$
$$
\begin{cases}
g(z,z)=ar^m & \text{for } z\in V_{m+1}\setminus V_m,\\
g(z,z')=br^m & \text{for } z,z'\in V_{m+1}\setminus V_m \text{ with } z,z'\in F_w K \text{ for } |w|=m, \text{ and } z\neq z'.
\end{cases}
$$

For simplicity write $G_M(x, y)$ for (2.6.12). We need to compute $(-\Delta_\mu)_x G_M$ (x, y). Of course, via the weak formulation, we actually work with

$$\mathcal{E}(G_M(\cdot, y), v) \quad \text{for } v \in \text{dom}_0 \mathcal{E}.$$

The analog of (2.6.11) will read

(2.6.17)
$$\mathcal{E}(G_M(\cdot, y), v) = \sum_{z \in V_{M+1} \setminus V_0} v(z) \psi_z^{(M+1)}(y).$$

If we can establish (2.6.17), then the rest is easy. We just multiply by $f(y)$ and integrate, using standard arguments to interchange the energy and integral, to obtain

(2.6.18)
$$\mathcal{E}(u_M, v) = \int_K f(y) \sum_{z \in V_{M+1} \setminus V_0} v(z) \psi_z^{(M+1)}(y) d\mu(y)$$

for

(2.6.19)
$$u_M(x) = \int G_M(x, y) f(y) d\mu(y).$$

More routine arguments show that

(2.6.20)
$$\sum_{z \in V_{M+1} \setminus V_0} v(z) \psi_z^{(M+1)}(y) \to v(y) \quad \text{uniformly as } M \to \infty,$$

so the right side of (2.6.18) converges to $\int_K f v d\mu$, while the left side converges to $\mathcal{E}(u, v)$ for

(2.6.21)
$$u(x) = \int_K G(x, y) f(y) d\mu(y).$$

This is exactly $-\Delta_\mu u = f$. Of course we could write (2.6.17) as

(2.6.22)
$$-(\Delta_\mu)_x G_M(x, y) = \sum_{z \in V_{M+1} \setminus V_0} \delta(z, x) \psi_z^{(M+1)}(y),$$

more in line with (2.6.11), and then deduce

(2.6.23)
$$-(\Delta_\mu)_x G(x, y) = \delta(x, y).$$

This is just shorthand for the above discussion.

So our goal is to establish (2.6.17). We will first do this for $M = 0$, and the self-similarity will take care of the general case. We have

$$G_0(x, y) = a \left(\psi_{z_0}^{(1)}(x) \psi_{z_0}^{(1)}(y) + \psi_{z_1}^{(1)}(x) \psi_{z_1}^{(1)}(y) + \psi_{z_2}^{(1)}(x) \psi_{z_2}^{(1)}(y) \right)$$
$$+ b \left(\psi_{z_0}^{(1)}(x) \psi_{z_1}^{(1)}(y) + \psi_{z_0}^{(1)}(x) \psi_{z_2}^{(1)}(y) + \psi_{z_1}^{(1)}(x) \psi_{z_0}^{(1)}(y) \right.$$
$$+ \left. \psi_{z_1}^{(1)}(x) \psi_{z_2}^{(1)}(y) + \psi_{z_2}^{(1)}(x) \psi_{z_0}^{(1)}(y) + \psi_{z_2}^{(1)}(x) \psi_{z_1}^{(1)}(y) \right),$$

where $\{z_0, z_1, z_2\} = V_1 \setminus V_0$. We also have

$$\mathcal{E}\left(\psi_{z_0}^{(1)}, v \right) = \mathcal{E}_1\left(\psi_{z_0}^{(1)}, v \right) = 4r^{-1} v(z_0) - r^{-1} v(z_1) - r^{-1} v(z_2),$$

and so on, so

$$\mathcal{E}(G_0(\cdot, y), v) = a(4r^{-1}v(z_0)\psi_{z_0}^{(1)}(y) - r^{-1}v(z_1)\psi_{z_0}^{(1)}(y) - r^{-1}v(z_2)\psi_{z_0}^{(1)}(y) + \cdots)$$
$$+ b(4r^{-1}v(z_0)\psi_{z_1}^{(1)}(y) - r^{-1}v(z_1)\psi_{z_1}^{(1)}(y)$$
$$- r^{-1}v(z_2)\psi_{z_1}^{(1)}(y) + \cdots).$$

When we collect all the terms involving $v(z_0)$ we obtain

$$a\left(4r^{-1}\psi_{z_0}^{(1)}(y) - r^{-1}\psi_{z_1}^{(1)}(y) - r^{-1}\psi_{z_2}^{(1)}(y)\right)$$
$$+ b\left(3r^{-1}\psi_{z_1}^{(1)}(y) + 3r^{-1}\psi_{z_2}^{(1)}(y) - 2r^{-1}\psi_{z_0}^{(1)}(y)\right)$$
$$= \left(4r^{-1}a - 2r^{-1}b\right)\psi_{z_0}^{(1)}(y) + (-r^{-1}a + 3r^{-1}b)(\psi_{z_1}^{(1)}(y) + \psi_{z_2}^{(1)}(y)).$$

Since we want $v(z_0)$ to multiply $\psi_{z_0}^{(1)}(y)$, we just have to solve the linear equations

(2.6.24)
$$\begin{cases} 4r^{-1}a - 2r^{-1}b = 1, \\ -r^{-1}a + 3r^{-1}b = 0. \end{cases}$$

Since $r = \frac{3}{5}$ the solution is

(2.6.25)
$$a = \frac{9}{50}, \qquad b = \frac{3}{50}.$$

With these choices we end up with

(2.6.26) $\qquad \mathcal{E}(G_0(\cdot, y)v) = v(z_0)\psi_{z_0}^{(1)}(y) + v(z_1)\psi_{z_1}^{(1)}(y) + v(z_2)\psi_{z_2}^{(1)}(y),$

which is (2.6.17) for $M = 0$.

We are now in a position to prove (2.6.17) by induction on M. For simplicity we just do the case $M = 1$. It is clear that $\mathcal{E}(G_1(\cdot, y), v)$ is a linear combination of $v(z)$ for $z \in V_2 \setminus V_0$, so we need to compute the coefficient of $v(z)$. Now for $z \in V_2 \setminus V_1$ the coefficient must be $\psi_z^{(2)}(y)$ by the same reasoning that led to (2.6.26), the only difference being that there is an additional factor of r^{-1} in the computation of $\mathcal{E}_2(\psi_z^{(2)}, v)$, and this cancels the factor of r in (2.6.15) (or (2.6.16)). We need a somewhat different argument for the points in $V_1 \setminus V_0$. Consider the point z_0 in Figure 2.6.2. There are many terms in G_1 that contribute to the coefficient of $v(z_0)$.

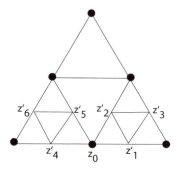

Figure 2.6.2

The terms in G_0 contribute $\psi_{z_0}^{(1)}(y)$, as we saw in (2.6.26). But there are also the terms that contain $\psi_{z_1'}^{(2)}(x)$, $\psi_{z_2'}^{(2)}(x)$, $\psi_{z_4'}^{(2)}(x)$, and $\psi_{z_5'}^{(2)}(x)$. For example,

$$r\mathcal{E}_2\left(\psi_{z_1'}^{(2)}, v\right) = \frac{20}{3}v\left(z_1'\right) - \frac{5}{3}v\left(z_0\right) - \frac{5}{3}v\left(z_2'\right) - \frac{5}{3}v\left(z_3'\right).$$

Since G_1 contains $r\psi_{z_1'}^{(2)}(x)$ multiplied by $\frac{9}{50}\psi_{z_1'}^{(2)}(y) + \frac{3}{50}\psi_{z_2'}^{(2)}(y) + \frac{3}{50}\psi_{z_3'}^{(2)}(y)$, the total contribution is

$$-\frac{3}{10}\psi_{z_1'}^{(2)}(y) - \frac{1}{10}\psi_{z_2'}^{(2)}(y) - \frac{1}{10}\psi_{z_3'}^{(2)}(y).$$

Summing the contributions from all four neighbors yields

$$-\frac{2}{5}\psi_{z_1'}^{(2)}(y) - \frac{2}{5}\psi_{z_2'}^{(2)}(y) - \frac{1}{5}\psi_{z_3'}^{(2)}(y),$$

$$-\frac{2}{5}\psi_{z_4'}^{(2)}(y) - \frac{2}{5}\psi_{z_5'}^{(2)}(y) - \frac{1}{5}\psi_{z_6'}^{(2)}(y).$$

When we add this to $\psi_{z_0}^{(1)}(y)$ we get a piecewise harmonic function which vanishes at $z_1', z_2', z_3', z_4', z_5', z_6'$ by the "$\frac{1}{5} - \frac{2}{5}$ rule", so the coefficient is exactly $\psi_{z_0}^{(2)}(y)$. This yields (2.6.17) for $M = 1$.

We now summarize what we have accomplished.

THEOREM 2.6.1 *On SG, the Dirichlet problem*

$$(2.6.27) \qquad\qquad -\Delta_\mu u = f, \qquad u\big|_{V_0} = 0$$

has a unique solution in dom Δ_μ *for any continuous f, given by*

$$(2.6.28) \qquad\qquad u(x) = \int_K G(x, y) f(y) d\mu(y)$$

for the Green's function

$$G(x, y) = \lim_{M \to \infty} G_M(x, y) \quad \text{(uniform limit)}$$

with G_M defined by (2.6.12), (2.6.16), and (2.6.25); or equivalently G is given by (2.6.15). In other words, if we denote by dom$_0 \Delta_\mu$ *the subspace of* dom Δ_μ *of functions vanishing on the boundary, then*

$$\Delta_\mu : \text{dom}_0 \Delta_\mu \to C(K)$$

is invertible, and the integral operator (2.6.28) is its inverse.

As in the case $K = I$, we see by inspection that the Green's function is positive, continuous, and symmetric. For each fixed y, $G(\cdot, y)$ is harmonic in the complement of the point y, meaning that its restriction to any cell not containing y in its interior is harmonic. In fact, if y is a junction point in V_m we may characterize $G(\cdot, y)$ as the unique continuous function harmonic away from y, vanishing at the boundary, and having its normal derivatives at y sum to 1. By making the values $G(x, y)$ with $x \in V_m \setminus V_0$ variables, and writing the harmonic condition $\Delta_m G(x, y) = 0$ at $x \neq y$ and the normal derivative condition at $x = y$ as linear equations in these variables, we obtain a system of equations whose unique solution gives those values; then G is determined by harmonic extension.

EXERCISES

2.6.1. Show that the inhomogeneous Dirichlet problem $-\Delta_\mu u = f$, $u(q_i) = a_i$, $i = 0, 1, 2$, has a unique solution

$$u(x) = \int_K G(x, y) f(y) d\mu(y) + h(x),$$

where h is the harmonic function satisfying $h(q_0) = a_i$, $i = 0, 1, 2$.

2.6.2. A function in dom Δ_μ is called *subharmonic* if $\Delta_\mu u \geq 0$. Show that a subharmonic function satisfies $u(x) \leq h(x)$, where h is the harmonic function satisfying $h(q_i) = u(q_i)$, $i = 0, 1, 2$.

2.6.3. Show that the Neumann problem $-\Delta_\mu u = f$, $\partial_n u(q_i) = 0$, $i = 0, 1, 2$, is solvable if and only if $\int_K f(x) d\mu(x) = 0$, and in that case the solution is unique up to an additive constant.

2.6.4. Show that the space $\mathcal{H}_k$ of solutions to $\Delta_\mu^{k+1} u = 0$ has dimension $3k + 3$.

2.6.5. Show that $G(x, y) \leq G(x, x)$.

2.6.6. Verify directly that (2.6.1) and (2.6.5) give the explicit solution to $-u'' = f$, $u(0) = u(1) = 0$.

2.6.7. Derive (2.6.18) from (2.6.17).

2.6.8. Prove (2.6.20) and use it to show that the right side of (2.6.18) converges to $\int_K f v d\mu$.

2.6.9. Prove that the left side of (2.6.18) converges to $\mathcal{E}(u, v)$ for u given by (2.6.21). In particular, show that u_M converges to u both uniformly and in energy.

2.6.10.* If $f \in L^2(d\mu)$, show that u defined by (2.6.28) belongs to $\text{dom}_{L^2} \Delta_\mu$ and $-\Delta_\mu u = f$.

2.6.11.* If there exists a function u in $\text{dom}_0 \mathcal{E}$ that minimizes $\frac{1}{2}\mathcal{E}(u, u) + \int_K f u d\mu$, show that u solves the Dirichlet problem (2.6.27). (Hint: Try $u + tv$ for $v \in \text{dom}_0 \mathcal{E}$.)

2.6.12. Show that $\partial_n G(q_i, y) = -h_i(y)$. (Hint: Apply Gauss–Green to $\int_K h_i \Delta u d\mu$.) Use this to show that u given by (2.6.28) satisfies $\partial_n u(q_i) = 0$, $i = 0, 1, 2$, if and only if $\int_K f h d\mu = 0$ for all harmonic functions h.

2.6.13.* Show that if u and f are continuous and $\int_K u \Delta v d\mu = \int f v d\mu$ for every $v \in \text{dom } \Delta_\mu$ with $v|_{V_0}$ and $\partial_n v|_{V_0}$ both zero, then $u \in \text{dom } \Delta_\mu$ and $\Delta_\mu u = f$. (Hint: Write $v(x) = \int_K G(x, y) g(y) d\mu(y)$ and use the results of Exercise 2.6.12.)

2.6.14.* For each word w define the 3×3 matrix G_w by $(G_w)_{jk} = G(F_w q_j, F_w q_k)$. Show the recursion formula

$$G_{wi} = A_i G_w A_i^* + r^{|w|} B_i,$$

where A_i are the harmonic extension matrices, given by (1.3.28), and

$$B_0 = \begin{pmatrix} 0 & 0 & 0 \\ 0 & \frac{9}{50} & \frac{3}{50} \\ 0 & \frac{3}{50} & \frac{9}{50} \end{pmatrix}, \quad B_1 = \begin{pmatrix} \frac{9}{50} & 0 & \frac{3}{50} \\ 0 & 0 & 0 \\ \frac{3}{50} & 0 & \frac{9}{50} \end{pmatrix}, \quad B_3 = \begin{pmatrix} \frac{9}{50} & \frac{3}{50} & 0 \\ \frac{3}{50} & \frac{9}{50} & 0 \\ 0 & 0 & 0 \end{pmatrix}.$$

2.6.15.* Use the fact that the Green's operator (2.6.21) is compact to conclude that there is an orthonormal basis in $L^2(d\mu)$ of Dirichlet eigenfunctions, with eigenvalues tending to infinity.

2.6.16.* Use the results of Exercise 2.6.15 to show that dom Δ_μ is dense in $L^2(d\mu)$.

2.6.17.* Show that there is an orthonormal basis in $L^2(d\mu)$ of Neumann eigenfunctions, with eigenvalues tending to infinity.

2.6.18.* Show that the Neumann eigenfunctions are the stationary points of the Rayleigh quotient $R(u) = \mathcal{E}(u)/\int_K u^2 d\mu$ on dom $\mathcal{E}$, while the Dirichlet eigenfunctions are the stationary points of the Rayleigh quotient on $\text{dom}_0 \mathcal{E}$.

2.7 LOCAL BEHAVIOR OF FUNCTIONS

We will first look at the local behavior of harmonic functions and then see to what extent the same behavior holds for functions in the domain of the standard Laplacian. In this section we work entirely on SG.

First we consider behavior near a boundary point, say q_0. We have already seen that there is a natural basis for $\mathcal{H}_0$ for this question, namely $h_0 + h_1 + h_2 = 1$, $h_1 + h_2$, and $h_1 - h_2$, since these are eigenfunctions for the zoom $h \to h \circ F_0$ at q_0. We are interested in the rate of convergence of h to $h(q_0)$ along the natural system of neighborhoods $F_0^m K$, so it is natural to look at

$$(2.7.1) \qquad\qquad \sup_{F_0^m K} |u(x) - u(q_0)| = \varepsilon_m.$$

Since harmonic functions satisfy the maximum principle, the value of ε_m is assumed when $x = F_0^m q_1$ or $F_0^m q_2$. Thus for $u \equiv 1$, $\varepsilon_m = 0$, for $u = h_1 + h_2$, $\varepsilon_m = (\frac{3}{5})^m$, and for $u = h_1 - h_2$, $\varepsilon_m = (\frac{1}{5})^m$. Also, we can recognize $h_1 + h_2$ by the condition that its normal derivative is nonzero, $\partial_n(h_1 + h_2)(q_0) = -2$. In other words, if u is harmonic and $\partial_n u(q_0) = 0$, then u is a linear combination of 1 and $h_1 - h_2$. Thus we have the following dichotomy.

LEMMA 2.7.1 *If u is harmonic and $\partial_n u(q_0) \neq 0$, then*

$$(2.7.2) \qquad\qquad c_1 \left(\frac{3}{5}\right)^m \leq \varepsilon_m \leq c_2 \left(\frac{3}{5}\right)^m,$$

while if $\partial_n u(q_0) = 0$, then

$$(2.7.3) \qquad\qquad \varepsilon_m = c \left(\frac{1}{5}\right)^m.$$

To complete the description we would like to have another local derivative at q_0 that reverses the roles of $h_1 + h_2$ and $h_1 - h_2$.

DEFINITION 2.7.2 *The tangential derivative $\partial_T u(q_0)$ is defined to be the limit*

$$(2.7.4) \qquad\qquad \partial_T u(q_0) = \lim_{m \to \infty} 5^m (u(F_0^m q_1) - u(F_0^m q_2))$$

if the limit exists.

While the normal derivative depends only on the symmetric part of u, the tangential derivative depends only on the skew-symmetric part. Since 1 and $h_1 + h_2$ are symmetric, they have tangential derivative equal to 0, while a simple computation shows $\partial_T(h_1 - h_2)(q_0) = 2$ and there is no need to take the limit. So the tangential derivative exists for harmonic functions, and the triple $(u(q_0), \partial_n u(q_0), \partial_T u(q_0))$, which may be thought of as a kind of first-order jet of u at q_0, uniquely determines the harmonic function. (In smooth analysis, the k-jet of a function at a point is the vector giving the value of the function and all derivatives up to order k at that point.)

THEOREM 2.7.3 *If u is harmonic, then u is uniquely determined by the 1-jet $(u(q_0), \partial_n u(q_0), \partial_T u(q_0))$ as follows:*

$$(2.7.5) \quad u(x) = u(q_0) - \frac{1}{2}\partial_n u(q_0)(h_1(x) + h_2(x)) + \frac{1}{2}\partial_T u(q_0)(h_1(x) - h_2(x)).$$

It is interesting that the factor 5 in the definition of the tangential derivative is greater than the factor $\frac{5}{3}$ in the definition of the normal derivative. Both factors are simply the reciprocals of the nontrivial eigenvalues of the matrices A_i. It might make sense to regard the tangential derivative as a "higher order" derivative than the normal derivative. At present, the tangential derivative plays a rather peripheral role in the theory, while the normal derivative plays a prominent role because of the Gauss–Green formula and the gluing principle.

It is possible to extend the definition of tangential derivative to junction points. Again there will be two distinct tangential derivatives, one with respect to each of the cells meeting at the junction point. We will not write down the formula, as the notation is a bit awkward. (For a sign convention we always move counterclockwise around the cells.) Unlike the case of normal derivatives, the two tangential derivatives at a junction point are unrelated. There are no matching conditions for tangential derivatives.

We will examine next the local behavior of harmonic functions near a junction point. To be specific, we take the point $x_0 = F_1 q_2 = F_2 q_1$ opposite q_0. If we restrict the harmonic function u to $F_1 K$ or $F_2 K$, we obtain a pair of harmonic functions $u \circ F_1$ and $u \circ F_2$, and the behavior of $u \circ F_1$ near q_2 and $u \circ F_2$ near q_1, when rescaled, gives the desired behavior of u near x_0. Each of these functions is determined by a triple of values, $(u \circ F_1(q_2), \partial_n(u \circ F_1)(q_2), \partial_T(u \circ F_1)(q_2))$ for the first and $(u \circ F_2(q_1), \partial_n(u \circ F_2)(q_1), \partial_T(u \circ F_2)(q_1))$ for the second. But these values are not independent, since we know $u \circ F_1(q_2) = u \circ F_2(q_1)$ and $\partial_n(u \circ F_1)(q_2) = -\partial_n(u \circ F_2)(q_1)$ by the matching conditions. This leaves us with four independent values. But this is one too many, since $\mathcal{H}_0$ is only three-dimensional! What is going on?

The explanation of this paradox is that we are really describing the local behavior near x_0 of functions harmonic on the subset $F_1 K \cup F_2 K$ of K. This subset has four boundary points, $F_1 q_0, q_1, q_2, F_2 q_0$, and so this space of harmonic functions is four-dimensional. It is a simple exercise to write down an invertible 4×4 matrix associating the values of u at these four boundary points with the four independent values $(u(F_1 q_2), \partial_n u(F_1 q_2), \partial_T u(F_1 q_2), \partial_T u(F_2 q_1))$. But not all these harmonic functions extend to global harmonic functions on K (it is easy to see that if an extension exists, then it is unique). Since there is a three-dimensional space of extensions, there must be a single linear obstruction to extension. We note that symmetric functions always extend, since there is a two-dimensional space of symmetric harmonic functions on both K and $F_1 K \cup F_2 K$. However, there is a two-dimensional space of skew-symmetric harmonic functions on $F_1 K \cup F_2 K$, but only a one-dimensional space of skew-symmetric harmonic functions on K. The secret lies in Exercise 1.4.7, in which a skew-symmetric function that is harmonic in the complement of q_0 and has a singularity at q_0 is constructed. This function restricts to a nice harmonic function on $F_1 K \cup F_2 K$ that does not extend.

This might seem like a satisfactory explanation, but there is more going on. If we take a different junction point, say in $V_2 \setminus V_1$, then there will still be a four-dimensional space of harmonic functions in the union of 2-cells meeting at the junction point, and there will still be a single linear obstruction to extension, but it will not be the same obstruction! Looked at another way, the local behavior of global harmonic functions at different junction points is not the same. This is the first appearance of a principle we will call *geography is destiny*. (Or, if you prefer, "you can tell where you are by looking at the scenery," "don't try to grow bananas on Manhattan island," . . . .) We will see other instances of it later on. We leave the detailed verification to the exercises.

We now consider a function in the domain of the standard Laplacian dom Δ. The situation is a bit more complicated, because we must either settle for weaker conclusions or make stronger hypotheses. First we prove a slightly weaker version of the dichotomy in Lemma 2.7.1.

THEOREM 2.7.4 *Let $u \in$ dom Δ. If $\partial_n u(q_0) \neq 0$ then*

$$(2.7.6) \qquad\qquad c_1 \left(\frac{3}{5}\right)^m \leq \varepsilon_m \leq c_2 \left(\frac{3}{5}\right)^m,$$

while if $\partial_n u(q_0) = 0$ then

$$(2.7.7) \qquad\qquad \varepsilon_m \leq cm5^{-m}.$$

In comparison with Lemma 2.7.1, we see that (2.7.6) is identical to (2.7.2), but (2.7.7) is slightly weaker than (2.7.3) because of the factor m. It is possible to show that the estimate is true without this factor, provided we assume that Δu satisfies a Hölder condition. We will need a few preliminary lemmas before we can prove the theorem, but first we give a surprising application.

COROLLARY 2.7.5 *Let u be any nonconstant function in dom Δ. Then u^2 is not in dom Δ.*

Proof: Let x_0 be a junction point where $\partial_n u(x_0) \neq 0$. We have already observed in Section 2.4 that such points exist. Write $u(x_0) = a$. Then $u = (u - a) + a$ and $u^2 = (u - a)^2 + 2a(u - a) + a^2$. Since $2a(u - a) + a^2$ is in dom Δ, it suffices to show that $(u - a)^2$ is not in dom Δ. Now we localize the theorem to x_0. Since $\partial_n (u - a)(x_0) \neq 0$ we have the estimate (2.7.6), and squaring it we obtain

$$(2.7.8) \qquad\qquad c_1^2 \left(\frac{3}{5}\right)^{2m} \leq \varepsilon_m' \leq c_2^2 \left(\frac{3}{5}\right)^{2m}$$

for ε'_m equal to the sup of $u - a$ in the m-cells containing x_0. Now it is clear that neither (2.7.6) nor (2.7.7) can hold for ε'_m, as the decay rate is too fast for (2.7.6) and too slow for (2.7.7) (here we use $(\frac{3}{5})^2 > \frac{1}{5}$). So $(u - a)^2 \notin \operatorname{dom} \Delta$. $\qquad\square$

It is easy to adapt the proof to show that $f(u) \notin \operatorname{dom} \Delta$ for any nonlinear differentiable function f. Similarly, if u and v are nonconstant functions in $\operatorname{dom} \Delta$, then uv is not in $\operatorname{dom} \Delta$. This is a stronger statement than saying $\operatorname{dom} \Delta$ does not form an algebra under pointwise multiplication. It actually says that multiplication is never good. Forget multiplying by cut-off functions. Forget partitions of unity. Many of the standard tools in PDE theory are just not going to be available on fractals.

The first lemma on the way to the proof of Theorem 2.7.4 is a quantitative statement about the convergence of $\frac{3}{2} 5^n \Delta_m u$ to Δu.

LEMMA 2.7.6 *Let $u \in \operatorname{dom} \Delta$ and let $x \in V_m \setminus V_0$. Then*

$$(2.7.9) \qquad \Delta u(x) - \frac{3}{2} 5^m \Delta_m u(x) = \frac{3}{2} 3^m \int \psi_x^{(m)}(y)(\Delta u(x) - \Delta u(y)) d\mu(y).$$

Proof: In the proof of Theorem 2.4.2 we found

$$\int \psi_x^{(m)}(y) \Delta u(y) d\mu(y) = \left(\frac{5}{3}\right)^m \Delta_m u(x)$$

(this was (2.4.7)). But we also know $\int \psi_x^{(m)}(y) d\mu(y) = \frac{2}{3^{m+1}}$ for the standard measure, and combining these we obtain (2.7.9). $\qquad\square$

Next we give a refinement of the "$\frac{1}{5} - \frac{2}{5}$ rule".

LEMMA 2.7.7 *Let $u \in \operatorname{dom} \Delta$; consider any $(m-1)$-cell with boundary vertices y_0, y_1, y_2; and let $x_0, x_1, x_2 \in V_m \setminus V_{m-1}$ be the vertex in that cell with x_j opposite y_j. Then*

$$u(x_2) = \frac{2}{5}(u(y_0) + u(y_1)) + \frac{1}{5} u(y_2)$$
$$(2.7.10)$$
$$+ \frac{2}{3} \frac{1}{5^m} \left(\frac{6}{5} \Delta u(x_2) + \frac{2}{5} \Delta u(x_1) + \frac{2}{5} \Delta u(x_0)\right) + R_m,$$

and so on, with

$$(2.7.11) \qquad\qquad\qquad R_m = o(5^{-m}).$$

Proof: We use (2.7.9) at each of the points x_j. Note that the right side of (2.7.9) is $o(1)$ because it is the average of the difference $\Delta u(x) - \Delta u(y)$ over a small region, and Δu is continuous. Thus if we multiply by $\frac{2}{3} \cdot 5^{-m}$ we obtain

$$(2.7.12) \qquad\qquad \Delta_m u(x_j) = \frac{2}{3} \cdot \frac{1}{5^m} \Delta u(x_j) + o(5^{-m}).$$

Now we solve the system of equations (2.7.12) for $u(x_j)$ and we obtain (2.7.10). We leave the details to the exercises. $\qquad\square$

Proof of Theorem 2.7.4: Without loss of generality assume $u(q_0) = 0$. For simplicity we prove the estimates for $u(F_0^m q_1)$ and $u(F_0^m q_2)$. (Note that this suffices for the proof of Corollary 2.7.5.) Recall from (2.4.9) that

$$(2.7.13) \qquad u(F_0^m q_1) + u(F_0^m q_2) = -\left(\frac{3}{5}\right)^m \partial_n u(q_0) + O(5^{-m}),$$

so the crux of the matter is to control the difference $u(F_0^m q_1) - u(F_0^m q_2)$. Now by subtracting (2.7.10) at the points $F_0^m q_1$ and $F_0^m q_2$ we obtain

$$(2.7.14) \qquad u\left(F_0^m q_1\right) - u\left(F_0^m q_2\right) = \frac{1}{5}\left(u\left(F_0^{m-1} q_1\right) - u\left(F_0^{m-1} q_2\right)\right) + O\left(5^{-m}\right).$$

This is a recursion relation for the difference, and it is easy to see that this implies

$$(2.7.15) \qquad |u(F_0^m q_1) - u(F_0^m q_2)| \le cm5^{-m}.$$

(This is the place in the argument where we pick up the unwanted factor of m.) If $\partial_n u(q_0) \ne 0$ then (2.7.13) and (2.7.15) together imply

$$(2.7.16) \qquad c_1\left(\frac{3}{5}\right)^m \le |u(F_0^m q_j)| \le c_2\left(\frac{3}{5}\right)^m$$

for $j = 1, 2$, as desired. On the other hand, if $\partial_n u(q_0) = 0$, then together they imply

$$(2.7.17) \qquad |u(F_0^m q_j)| \le cm5^{-m} \quad \text{for } j = 1, 2, \text{ as desired.} \qquad \square$$

Note that (2.7.14) is very suggestive of the existence of a tangential derivative, but is just shy of proving it. Indeed, we need to assume some smoothness for Δu before we can prove the existence of tangential derivatives.

THEOREM 2.7.8 *Assume $u \in \text{dom } \Delta$ and Δu satisfies a Hölder condition of some order. Then $\partial_T u(q_0)$ exists.*

Proof: Let $\tilde{h}$ denote the skew-symmetric harmonic function with singularity at q_0 described in Exercise 1.4.7, with $\tilde{h}(F_0^k q_1) = -\tilde{h}(F_0^k q_2) = 3^k$. Let $\tilde{h}_m$ denote the piecewise harmonic function that agrees with $\tilde{h}$ on $K \setminus F_0^m K$, and on $F_0^m K$ has boundary values $\tilde{h}_m(F_0^m q_1) = -\tilde{h}_m(F_0 q_2) = 3^m$ and $\tilde{h}_m(q_0) = 0$. Note that $\tilde{h}_m$ is also skew-symmetric, so $\partial_n \tilde{h}_m(q_0) = 0$. Now $\tilde{h}$ is not integrable. In fact it is easy to see that $\int |\tilde{h}_m| d\mu = O(m)$. But the fact that $\Delta u = f$ is Hölder continuous implies that the improper integral

$$(2.7.18) \qquad \int \tilde{h} f d\mu = \lim_{m \to \infty} \int \tilde{h}_m f d\mu$$

exists. Indeed, by the skew-symmetry,

$$(2.7.19) \qquad \int \tilde{h}_m(y) f(y) d\mu(y) = \frac{1}{2} \int \tilde{h}_m(y)(f(y) - f(Ry)) d\mu(y),$$

where R denotes the reflection preserving q_0, and the Hölder estimate for $f(y) - f(Ry)$ suffices to establish (2.7.18). By the way, the contribution to $\int \tilde{h}_m f d\mu$ coming from $F_0^m K$ tends to zero in the limit.

Now we apply the Gauss–Green formula to the functions u and $\tilde{h}$. More precisely, we apply the local version to $K \setminus F_0^m K$ and $F_0^m K$ separately, and then add. So there is no contribution from the $u \Delta \tilde{h}$ term, and also the sum of the $\tilde{h} \partial_n u$ terms at $F_0^m q_j$, $j = 1, 2$, will vanish by the matching conditions for $\partial_n u$. Thus we find

$$(2.7.20) \quad \int \tilde{h}_m f \, d\mu = -u(q_1) \partial \tilde{h}(q_1) - u(q_2) \partial \tilde{h}(q_2) - b_m (u(F_0^m q_1) - u(F_0^m q_2)),$$

where b_m is the sum of the normal derivatives of $\tilde{h}_m$ at $F_0^m q_1$. We can easily compute b_m by taking these normal derivatives at the m-cell level, since

$$\tilde{h}_m \left(F_0^m q_1 \right) = -\tilde{h}_m \left(F_0^m q_2 \right) = 3^m, \qquad \tilde{h}_m \left(F_0^{m-1} q_1 \right) = 3^{m-1},$$

and $\tilde{h}_m(q_0) = \tilde{h}_m(F_0^{m-1} F_1 q_2) = 0$ by skew-symmetry, so

$$(2.7.21) \qquad b_m = \left(\frac{5}{3} \right)^m \left(4 \cdot 3^m - (-3^m) - 3^{m-1} \right) = \frac{14}{3} \cdot 5^m.$$

When we substitute (2.7.21) in (2.7.20) and take the limit we obtain

$$\partial_T u(q_0) = -\frac{5}{14} (u(q_1) - u(q_2)) - \frac{3}{14} \int \tilde{h} f \, d\mu. \qquad \square$$

The proof actually shows that the existence of the tangential derivative is equivalent to the existence of the limit (2.7.18). It is easy enough to construct continuous functions f for which this limit fails to exist, so by taking $u(x) = \int G(x, y) f(y) \, d\mu(y)$ for such functions we obtain examples of functions in dom Δ for which the tangential derivative does not exist.

Now suppose u satisfies the hypotheses of Theorem 2.7.8, and let h be the harmonic function with the same 1-jet as u at q_0. We may regard h as a *tangent* to u at q_0. The remainder $u - h = r$ satisfies the decay estimate

$$(2.7.22) \qquad |r \left(F_0^m q_j \right)| = O \left(5^{-m} \right) \quad \text{for } j = 1, 2,$$

but we get a slightly better estimate for the odd part, namely

$$(2.7.23) \qquad |r \left(F_0^m q_1 \right) - r \left(F_0^m q_2 \right)| = o \left(5^{-m} \right).$$

Indeed (2.7.23) is just a rephrasing of the condition $\partial_T r(q_0) = 0$, and then (2.7.22) follows from (2.7.13) since $r(q_0) = \partial_n r(q_0) = 0$. Similar reasoning shows that we can improve (2.7.17) to remove the factor m.

Finally, we show that the estimates we obtained at the boundary points of a cell transfer to the entire cell. This is a completely generic argument. Suppose $u \in \text{dom } \Delta$ with $\Delta u = f$ and let $F_w K$ be an m-cell, $|w| = m$. Suppose we know

$$(2.7.24) \qquad |u(F_w q_i)| \le a \quad \text{for } i = 0, 1, 2.$$

What can we say about the maximum value of $|u|$ on $F_w K$? Recall that

$$\Delta(u \circ F_w) = 5^{-m} f \circ F_w.$$

We write $u \circ F_w = h + g$, where h is the harmonic function taking the same boundary values as $u \circ F_w$, and so

(2.7.25) $$g(x) = -5^{-m} \int G(x, y) f(F_w y) d\mu(y).$$

Now $|h| \leq a$ by the maximum principle for harmonic functions, while (2.7.25) implies

(2.7.26) $$|g(x)| \leq c_0 \|f\|_\infty 5^{-m},$$

where c_0 is the constant

$$\max_y \int G(x, y) d\mu(y).$$

So we obtain

(2.7.27) $$|u| \leq a + c_0 \|f\|_\infty 5^{-m} \quad \text{on } F_w K.$$

This argument shows that the estimates (2.7.6) and (2.7.7) stated in Theorem 2.7.4 follow from (2.7.16) and (2.7.17) established in the proof. Similarly, (2.7.22) yields

(2.7.28) $$r\big|_{F_0^m K} = O(5^{-m}).$$

More careful reasoning leads from (2.7.23) to a statement that the odd part of r is $o(5^{-m})$.

EXERCISES

2.7.1. Show that for a harmonic function, the sum of the tangential derivatives at the boundary points is zero.

2.7.2. Find a 3×3 matrix that describes the transformation from the 1-jet of a harmonic function at q_0 to the 1-jet at q_1.

2.7.3. Find a 4×4 matrix describing the transformation from the values of u at $F_1 q_0, q_1, q_2, F_2 q_0$ to the values $(u(F_1 q_2), \partial_n u(F_1 q_2), \partial_T u(F_1 q_2), \partial_T u (F_2 q_1))$ for u harmonic in $F_1 K \cup F_2 K$. Show that the matrix is invertible.

2.7.4. (a) Find the linear equation in $(u(F_1 q_2), \partial_n u(F_1 q_2), \partial_T (F_1 q_2), \partial_T (F_2 q_1))$ that is satisfied if u is a global harmonic function. (b) Do the same at the point $F_1 F_1 q_2$.

2.7.5. Show that if u and v are nonconstant functions in dom Δ, then uv is not in dom Δ.

2.7.6. Show that if f is a nonlinear differentiable function and u is a nonconstant function in dom Δ, then $f(u)$ is not in dom Δ. Similarly, if u is also bounded away from zero, then $1/u$ is not in dom Δ.

2.7.7. Show that the solution of the system (2.7.12) yields (2.7.10).

2.7.8. Show that (2.7.14) implies (2.7.15).

2.7.9. Show that the limit (2.7.18) exists and the integral over $F_0^m K$ does not contribute to the limit if f satisfies a Hölder condition of any positive order.

2.7.10. Show that we can remove the factor m in (2.7.17) if we assume Δu satisfies a Hölder condition.

2.7.11. Show that

$$\max_{x \in F_0^m K} |r(x) - r(Rx)| = o(5^{-m}).$$

2.7.12.* Give an alternative proof of Theorem 2.7.8 by writing $u(x) = h(x) - \int G(x, y) f(y) d\mu(y)$, where h is harmonic, and estimating the size of $G(F_0^m q_1, y) - G(F_0^m q_2, y)$.

2.8 NOTES AND REFERENCES

Most of the material in Sections 2.1–2.4 and 2.6 is from [Kigami 1989]. We have chosen to put the weak formulation first. Theorem 2.3.3 on matching conditions for normal derivatives and the gluing results in Section 2.5 were first emphasized in [Strichartz 1999a]. The construction of the Green's function for I in Section 2.6 has not been published before. The dichotomy for harmonic functions (Lemma 2.7.1) was first observed in [Dalrymple et al. 1999]. The definition of tangential derivatives and the existence Theorem 2.7.8 are from [Strichartz 2000a], as is the definition of a tangent (see also [Teplyaev 2000]). The dichotomy Theorem 2.7.4 for functions in dom Δ and its proof are from [Ben-Bassat et al. 1999] (that reference has a typo in the statement of Lemma 2.7.7, which is corrected here). The proof of Corollary 2.7.5 is one of two from [Ben-Bassat et al. 1999]. The other proof is more informative in that it shows that $\Delta(u^2)$ really exists as a measure, but this measure is singular. The idea is that the formal identity $\Delta(u^2) = 2\nabla u \cdot \nabla u + 2u \Delta u$ makes sense if you identify $\nabla u \cdot \nabla u$ with the energy measure, and this is singular by [Kusuoka 1989] (a second proof of Kusuoka's result is also given).

The spaces $\mathcal{H}_k$ in Exercise 2.6.4 were studied in [Strichartz & Usher 2000], where they were used to construct spline spaces. The formula in Exercise 2.6.14 is from [Kigami et al. 2000], where it was called the "near diagonal formula."

Chapter Three

Spectrum of the Laplacian

3.1 FOURIER SERIES REVISITED

The Fourier series on I, sine or cosine, gives the expansion of a general function in terms of eigenfunctions, Dirichlet or Neumann, of the Laplacian. By convention, we write

$$(3.1.1) \qquad -\Delta u = \lambda u$$

for the eigenfunction equation. We happen to know all solutions to (3.1.1) for all complex λ, without specifying boundary conditions, namely $e^{\alpha x}$ where $-\alpha^2 = \lambda$, although we will eventually arrive at an alternate description using the self-similar structure. Of course, when we impose Dirichlet boundary conditions $u(0) = u(1) = 0$ we obtain the eigenfunctions $\{\sin \pi k x\}$, $k = 1, 2, \ldots$, with eigenvalues $\pi^2 k^2$, while if we impose Neumann boundary conditions $\partial_n u(0) = \partial_n u(1) = 0$, we obtain the eigenfunctions $\{\cos \pi k x\}$, $k = 0, 1, \ldots$, with eigenvalues $\pi^2 k^2$. By coincidence, the Dirichlet and Neumann spectra are identical, except for the Neumann eigenvalue 0. In fact, there are general principles that imply that the two spectra must be close, but the coincidence of eigenvalues must be attributed to coincidence.

We would like to understand these eigenfunctions and eigenvalues in terms of the sequence of graphs Γ_m approximating I. The key observation is that eigenfunctions of the continuous Laplacian on I, when restricted to V_m, are still eigenfunctions of the discrete Laplacian on Γ_m, but the eigenvalues are not the same:

$$
\begin{aligned}
-\Delta_m u(x) &= 2u(x) - u\left(x + \frac{1}{2^m}\right) - u\left(x - \frac{1}{2^m}\right) \\
(3.1.2) \qquad &= \left(2 - e^{\alpha/2^m} - e^{-\alpha/2^m}\right) u(x) \\
&= -4 \sinh^2 \frac{\alpha}{2^{m+1}} u(x)
\end{aligned}
$$

if $u(x) = e^{\alpha x}$, or

$$(3.1.3) \qquad -\Delta_m u(x) = 4 \sin^2 \frac{\pi k}{2^{m+1}} u(x)$$

if $u(x) = \sin \pi k x$ or $\cos \pi k x$. Note that the eigenvalue in (3.1.2) is unchanged if we replace α by $-\alpha$, so it depends only on the eigenvalue λ. We can represent this as

$$(3.1.4) \qquad -\Delta_m u \big|_{V_m} = \lambda_m u$$

for a sequence $\{\lambda_m\}$ of discrete eigenvalues depending on λ. We note that

$$(3.1.5) \qquad \lim_{m \to \infty} 4^m \lambda_m = \lambda,$$

as would be expected since

(3.1.6) $$\Delta = \lim_{m \to \infty} 4^m \Delta_m.$$

The expression for λ_m as a function of λ is easily read off from (3.1.2):

(3.1.7) $$\lambda_m = 4 \sin^2 \frac{\sqrt{\lambda}}{2^{m+1}}.$$

It is a transcendental function, and by coincidence it is easily expressible in terms of well-known functions. What is more important to us is the relationship between λ_m and λ_{m-1}. We compute

$$\lambda_{m-1} = 4 \sin^2 2 \frac{\sqrt{\lambda}}{2^{m+1}} = 4 \left(2 \sin \frac{\sqrt{\lambda}}{2^{m+1}} \cos \frac{\sqrt{\lambda}}{2^{m+1}} \right)^2$$

$$= 4 \sin^2 \frac{\sqrt{\lambda}}{2^{m+1}} \left(4 - 4 \sin^2 \frac{\sqrt{\lambda}}{2^{m+1}} \right),$$

or

(3.1.8) $$\lambda_{m-1} = \lambda_m (4 - \lambda_m).$$

We also need to solve for λ_m in terms of λ_{m-1}, which we write as

(3.1.9) $$\lambda_m = 2 + \varepsilon_m \sqrt{4 - \lambda_{m-1}}, \qquad \varepsilon_m = \pm 1.$$

Note that these relationships are algebraic. So, if we specify λ_{m_0} for some value m_0, then (3.1.8) specifies λ_m for all $m < m_0$. However, for $m > m_0$, we have the free choices of ε_m in (3.1.9) that allow infinitely many continuations. There is one condition we must impose, however. In order for the limit in (3.1.5) to exist, we need $\lambda_m \to 0$ as $m \to \infty$, and this means that all but a finite number of the ε_m must be chosen to be -1. Since

(3.1.10) $$2 - \sqrt{4 - x} = \frac{1}{4} x + O(x^2) \quad \text{as } x \to 0,$$

it is easy to see that the limit (3.1.5) exists in that case.

There is one small exception to the above description, namely that if $u|_{V_m} \equiv 0$ then we cannot claim that it is an eigenfunction. Does this actually happen? Yes, in fact, in exactly the cases in which we are most interested (for example, $\sin \pi 2^m x$ on V_m). However, any fixed eigenfunction has at most a finite number of zeroes in I, so for m large enough this will not happen. We define the *generation of birth*, m_0, to be the smallest m such that the restriction of u to $V_m \setminus V_0$ is not identically zero. For a generic eigenfunction we have $m_0 = 1$, and the eigenvalue equation on V_1 is the single equation

(3.1.11) $$2u \left(\frac{1}{2} \right) - u(0) - u(1) = \lambda_1 u \left(\frac{1}{2} \right), \qquad u \left(\frac{1}{2} \right) \neq 0.$$

We may regard this equation as giving the value λ_1 in terms of $u|_{V_1}$, but it also gives the value of $u(\frac{1}{2})$ in terms of the boundary values $u(0)$, $u(1)$, as

(3.1.12) $$u \left(\frac{1}{2} \right) = \frac{1}{2 - \lambda_1} (u(0) + u(1)) \quad \text{provided } \lambda_1 \neq 2.$$

We will therefore refer to 2 as the *forbidden* eigenvalue. Note that (3.1.12) is a variant of the harmonic extension algorithm (1.3.8), which corresponds to the case $\lambda_1 = 0$. What (3.1.12) says is that if we know $u\big|_{V_0}$ and the eigenvalue λ_1, then this determines $u\big|_{V_1}$ provided λ_1 is not forbidden.

The story repeats at all levels. That is, if we know $u\big|_{V_{m-1}}$ and λ_m, then this determines $u\big|_{V_m}$. Indeed, we only have to find the values $u(x)$ for $x \in V_m \setminus V_{m-1}$. Any such x lies between two consecutive points y_0, y_1 in V_{m-1}. The eigenvalue equation at x is

(3.1.13) $$2u(x) - u(y_0) - u(y_1) = \lambda_m u(x),$$

so we obtain

(3.1.14) $$u(x) = \frac{1}{2 - \lambda_m}(u(y_0) + u(y_1)), \quad \text{provided } \lambda_m \neq 2.$$

So now we have a recipe for both eigenvalues and eigenfunctions. Choose a value of λ_1 and $\{\varepsilon_m\}$ such that all but a finite number of $\varepsilon_m = -1$; determine λ_m for $m > 1$ by (3.1.9), and λ by (3.1.5). Assume that we never encounter the forbidden eigenvalue. Then u is determined on V_* inductively by (3.1.14) for any choice of boundary values $u(0)$ and $u(1)$, and then by continuity on all of I. We call this recipe *spectral decimation*.

We derived spectral decimation by starting with a knowledge of what the continuous eigenfunctions actually are. Now we want to see that it can stand alone.

LEMMA 3.1.1 *Suppose u_{m-1} defined on V_{m-1} satisfies the eigenvalue equation*

(3.1.15) $$-\Delta_{m-1} u_{m-1} = \lambda_{m-1} u_{m-1} \quad \text{on } V_{m-1} \setminus V_0.$$

Let λ_m be related to λ_{m-1} by (3.1.8) or (3.1.9), and assume $\lambda_m \neq 2$. Then extend u_{m-1} to u_m on V_m using (3.1.14). We obtain a function on V_m satisfying the eigenvalue equation

(3.1.16) $$-\Delta_m u_m = \lambda_m u_m \quad \text{on } V_m \setminus V_0.$$

Proof: The fact that the eigenvalue equation (3.1.16) holds on $V_m \setminus V_{m-1}$ follows from the equivalence of (3.1.14) and (3.1.13). What is not obvious, and is somewhat miraculous, is that it also holds on $V_{m-1} \setminus V_0$. To see this we consider three consecutive points y_0, y_1, y_2 in V_{m-1} and fill in to get five consecutive points y_0, x, y_1, z, y_2 in V_m (see Figure 3.1.1).

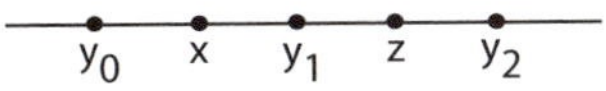

Figure 3.1.1

What we know is that

(3.1.17) $$(2 - \lambda_{m-1})u_{m-1}(y_1) = u_{m-1}(y_0) + u_{m-1}(y_2),$$

because this is (3.1.15) at y_1. What we need to show is that

(3.1.18) $$(2 - \lambda_m)u_{m-1}(y_1) = u_m(x) + u_m(z),$$

because this is (3.1.16) at y_1. Moreover, $u_m(x)$ and $u_m(z)$ are given by (3.1.14) as

$$u_m(x) = \frac{1}{2 - \lambda_m}(u_{m-1}(y_0) + u_{m-1}(y_1)),$$

$$u_m(z) = \frac{1}{2 - \lambda_m}(u_{m-1}(y_1) + u_{m-1}(y_2)),$$

so adding these yields

$$(3.1.19) \qquad u_m(x) + u_m(z) = \frac{1}{2 - \lambda_m}(u_{m-1}(y_0) + 2u_{m-1}(y_1) + u_{m-1}(y_2)).$$

Now we may substitute (3.1.17) into (3.1.19) to eliminate $u_{m-1}(y_0) + u_{m-1}(y_2)$, obtaining $u_m(x) + u_m(y) = \frac{2 + 2 - \lambda_{m-1}}{2 - \lambda_m} u_{m-1}(y_1)$. This will be the same as (3.1.18) provided $2 - \lambda_m = \frac{4 - \lambda_{m-1}}{2 - \lambda_m}$, and a little algebra shows that this is the same as (3.1.8). $\qquad\square$

The gist of the argument is that the relationship (3.1.8) between the eigenvalues emerges from the computation. That is, the extension algorithm (3.1.14) is equivalent to the eigenvalue equation at the new points, so this choice is dictated by λ_m, but then the eigenvalue equation at the old points dictates what λ_m must be. If we hadn't known (3.1.8) from the beginning, we could have discovered it in the course of the proof.

There are some technical details needed to complete the story, such as the fact that this recipe produces a continuous function u on V_*, and $-\Delta u = \lambda u$, but these are straightforward and are left to the exercise.

We now come to the description of what happens in the nongeneric cases, when the forbidden eigenvalue may appear and the generation of birth may be greater than 1. We first deal with the case of Dirichlet eigenfunctions, so $u(0) = u(1) = 0$, and $\#(V_m \setminus V_0) = 2^m - 1$. In this case we may regard Δ_m as a symmetric matrix of order $2^m - 1$, so it must have $2^m - 1$ real eigenvalues. When $m = 1$ we see by inspection that $u_1(\frac{1}{2}) = 1$ (the only choice, up to a constant multiple) produces an eigenfunction with eigenvalue $\lambda_1 = 2$, the forbidden value. (A good thing, too, because $u_1(\frac{1}{2})$ is not determined by $u(0)$ and $u(1)$.) So the spectrum of Δ_1 is $\{2\}$. Also, the eigenfunction is the restriction to V_1 of $\pm \sin \pi k x$ for any odd value of k. Passing to Δ_2, we have two choices of λ_2 corresponding to $\lambda_1 = 2$ (namely $\lambda_2 = \varphi_\pm(2)$ for

$$(3.1.20) \qquad\qquad \varphi_\pm(x) = 2 \pm \sqrt{4 - x}$$

by (3.1.9)) that are eigenvalues by Lemma 3.1.1. These eigenvalues have generation of birth $m_0 = 1$. We are still missing one eigenvalue, which must have $m_0 = 2$, so the eigenfunction must have $u(\frac{1}{2}) = 0$. By inspection we see that the choice $u(\frac{1}{4}) = 1$, $u(\frac{3}{4}) = -1$ produces an eigenfunction with $\lambda_2 = 2$. The spectrum of Δ_2, in increasing order, is thus $\{\varphi_-(2), 2, \varphi_+(2)\}$. Moreover, the associated eigenfunction is obtained by restricting $\sin \pi k x$ to V_2, for $k \equiv 1, 2, 3 \mod 4$.

The pattern continues. Having found the spectrum $\{\alpha_1, \ldots, \alpha_{2^{m-1}-1}\}$ and associated eigenfunctions on V_{m-1}, we may use Lemma 3.1.1 to produce $2^m - 2$

eigenvalues, $\{\varphi_{\pm}(\alpha_j)\}$. (We can check that these are distinct and not the forbidden value.) These have $m_0 < m$. We need one more eigenvalue with $m_0 = m$, and we easily check that we may take $\lambda_m = 2$ and the eigenfunction alternates ± 1 on the points in $V_m \setminus V_{m-1}$. Thus the spectrum of Δ_m in increasing order is

$$\{\varphi_-(\alpha_1), \ldots, \varphi_-(\alpha_{2^{m-1}-1}), 2, \varphi_+(\alpha_{2^{m-1}-1}), \ldots, \varphi_+(\alpha_1)\},$$

because φ_- preserves order and φ_+ reverses order. Moreover, the eigenfunctions are the restriction to V_m of $\sin \pi k x$ for $k \equiv 1, 2, \ldots, 2^m - 1 \mod 2^m$. The relationship between m_0 and the values of $\varepsilon_{m_0+1}, \ldots, \varepsilon_m$, and the value of k, is a bit complicated, so we leave the details to the exercises.

One consequence of the above discussion is the fact that $\sin \pi x$ takes on algebraic values for $x \in V_*$. In fact, the same is true for any rational value of x. Another consequence is that the ground state eigenfunction $\sin \pi x$ is positive and concave down, essentially because the coefficient $\frac{1}{2-\lambda_m}$ in (3.1.14) is greater than $\frac{1}{2}$.

We now briefly discuss the modifications needed to handle the Neumann eigenfunctions. It is not a priori clear what the Neumann conditions should be in the discrete case. The best way to think about it is that the function may be reflected evenly about each boundary point, and then will satisfy the eigenvalue equation at the boundary points as well. This is certainly true of the continuous eigenfunctions $\cos \pi k x$. Also, the analogous statement for Dirichlet eigenfunctions and odd reflection is trivially true. So this means that

$$(3.1.21) \qquad \begin{cases} 2u(0) - 2u\left(\frac{1}{2^m}\right) = \lambda_m u(0), \\ 2u(1) - 2u\left(1 - \frac{1}{2^m}\right) = \lambda_m u(1) \end{cases}$$

must be adjoined to the eigenvalue equations, which means that for Δ_m we have a matrix of order $2^m + 1$. When $m = 1$ we see by inspection that the eigenvalues are 0, 2, 4 with corresponding eigenfunctions $(u(0), u(\frac{1}{2}), u(1))$ given by $(1, 1, 1)$, $(1, 0, -1)$, and $(1, -1, 1)$. Passing to $m = 2$ we have no difficulty using Lemma 3.1.1 for $\lambda_1 = 0$ (whence $\lambda_2 = 0$ or 4) or $\lambda_1 = 2$ to produce four eigenfunctions. The eigenvalue $\lambda_1 = 4$ presents only one choice for λ_2, since $\varphi_{\pm}(4) = 2$, the forbidden eigenvalue. In this case (3.1.14) gives the indeterminate value $0/0$, but in fact if we interpret it as 0, then we obtain an eigenfunction with eigenvalue 2. In general, of the $2^{m-1} + 1$ eigenvalues of Δ_{m-1}, all but $\lambda_{m-1} = 4$ bifurcate and generate two values for λ_m, while $\lambda_{m-1} = 4$ extends to only the single eigenvalue $\lambda_m = 2$, again interpreting the right side of (3.1.14) as 0.

In summary, spectral decimation allows us to compute all eigenfunctions of Δ as limits of eigenfunctions of Δ_m, and these eigenfunctions may also be found explicitly. Also, the Dirichlet and Neumann spectrum of Δ is the limit of 4^m times the same spectrum of Δ_m. The orthogonality of Dirichlet or Neumann eigenfunctions corresponding to different eigenvalues follows from the symmetry of the Laplacians; in the discrete case the inner product is just

$$(3.1.22) \qquad \langle f, g \rangle_m = \sum_{x \in V_m \setminus V_0} f(x) g(x) + \frac{1}{2} \sum_{x \in V_0} f(x) g(x).$$

In order to give the correct formulas for Fourier coefficients we also need to compute the inner products

$$(3.1.23) \qquad \langle u, u \rangle = \int_0^1 |u(x)|^2 dx = \lim_{m \to \infty} \frac{1}{2^m} \langle u, u \rangle_m$$

for the eigenfunctions. This can also be done using spectral decimation to relate the values of $\langle u, u \rangle_{m-1}$ and $\langle u, u \rangle_m$ and then pass to the limit, but we leave the details to the exercises.

EXERCISES

3.1.1. Verify (3.1.3) using trigonometric double angle formulas.

3.1.2. Show that the extension algorithm (3.1.14) always leads to a uniformly continuous function on V_* provided all but a finite number of ε_m equal -1. Also show $-\Delta u = \lambda u$ if u is extended by continuity to I.

3.1.3. Show that $\sin \pi x$ is algebraic if x is rational.

3.1.4.* Find an algorithm that finds m_0 and the values of ε_m for $m > m_0$ in terms of the integer k (specifically its binary expansion) for the function $\sin \pi k x$.

3.1.5. Show that $\sin \pi x$ is positive and concave down on I, using only (3.1.14).

3.1.6.* Find the relationship between the values $\langle u, u \rangle_{m-1}$ and $\langle u, u \rangle_m$ by using (3.1.14). Use this to give an infinite product formula for $\langle u, u \rangle$.

3.2 SPECTRAL DECIMATION

Our goal in this section is to begin to duplicate the spectral decimation recipe on SG for the eigenfunctions of the standard Laplacian. Here we deal with generic eigenvalues, and in the next section we look at those corresponding to Dirichlet or Neumann boundary conditions. Our approach will be to obtain solutions of the eigenvalue equation

$$(3.2.1) \qquad -\Delta u = \lambda u \quad \text{on } K$$

as limits of solutions of the discrete version

$$(3.2.2) \qquad -\Delta_m u_m = \lambda_m u_m \quad \text{on } V_m \setminus V_0.$$

As in the case of the interval, we will be lucky that we may take $u_m = u|_{V_m}$. We should emphasize that there is no a priori reason this should hold. In fact, it is not true for other Laplacians Δ_μ on SG (or even on I), and there are some very symmetric fractals where nothing like this is true even for the standard Laplacian. But, as always, mathematicans shamelessly exploit good luck!

We look for the analog of Lemma 3.1.1: Given an eigenfunction u_{m-1} on V_{m-1} with eigenvalue λ_{m-1}, how can we extend it to u_m on V_m, so as to be an eigenfunction with eigenvalue λ_m, and what is the relationship between the two eigenvalues? We will essentially go back to the computations in Section 1.3 and redo them, adding in the eigenvalues. So consider an $(m-1)$-cell with boundary points x_0, x_1, x_2 and let y_0, y_1, y_2 denote the points in $V_m \setminus V_{m-1}$ in that cell, with y_i opposite x_i. We know the values $u(x_i)$, and we want to determine the values $u(y_j)$ (for simplicity of notation, we drop the subscripts on u). The λ_m-eigenvalue equation at the points $\{y_i\}$ gives us the system of equations

(3.2.3)
$$\begin{cases} (4-\lambda_m)u(y_0) = u(y_1) + u(y_2) + u(x_1) + u(x_2), \\ (4-\lambda_m)u(y_1) = u(y_0) + u(y_2) + u(x_0) + u(x_2), \\ (4-\lambda_m)u(y_2) = u(y_1) + u(y_0) + u(x_1) + u(x_0). \end{cases}$$

To solve, we add them and rearrange terms:

(3.2.4) $$(2-\lambda_m)(u(y_0) + u(y_1) + u(y_2)) = 2(u(x_0) + u(x_1) + u(x_2)).$$

We see already that 2 should be a forbidden eigenvalue, so we assume $\lambda_m \neq 2$ and obtain

(3.2.5) $$u(y_0) + u(y_1) + u(y_2) = \left(\frac{2}{2-\lambda_m}\right)(u(x_0) + u(x_1) + u(x_2)).$$

We add $u(y_i)$ to both sides of the corresponding equation in (3.2.3) and then use (3.2.5) to simplify the right side:

(3.2.6)
$$\begin{cases} (5-\lambda_m)u(y_0) = \left(\frac{2}{2-\lambda_m}\right)(u(x_0) + u(x_1) + u(x_2)) + u(x_1) + u(x_2), \\ (5-\lambda_m)u(y_1) = \left(\frac{2}{2-\lambda_m}\right)(u(x_0) + u(x_1) + u(x_2)) + u(x_0) + u(x_2), \\ (5-\lambda_m)u(y_2) = \left(\frac{2}{2-\lambda_m}\right)(u(x_0) + u(x_1) + u(x_2)) + u(x_1) + u(x_0). \end{cases}$$

Now we see that 5 should also be a forbidden eigenvalue, so we assume $\lambda_m \neq 5$ and obtain

(3.2.7) $$u(y_0) = \frac{(4-\lambda_m)(u(x_1) + u(x_2)) + 2u(x_0)}{(2-\lambda_m)(5-\lambda_m)}, \quad \text{etc.}$$

This is the analog of (3.1.14). Note that it is a variant of the "$\frac{1}{5} - \frac{2}{5}$ rule", because the coefficients $\frac{4-\lambda_m}{(2-\lambda_m)(5-\lambda_m)}$ and $\frac{2}{(2-\lambda_m)(5-\lambda_m)}$ corresponding to the adjacent and opposite vertices reduce to $\frac{2}{5}$ and $\frac{1}{5}$ when $\lambda_m = 0$.

Next, we go back to the points in $V_{m-1} \setminus V_0$ and compare the λ_{m-1}-eigenvalue equation on V_{m-1}, which we know holds, and the λ_m-eigenvalue equation on V_m, which we want to hold. So consider a point x_0 in $V_{m-1} \setminus V_0$. In addition to the $(m-1)$-cell considered above, there is another $(m-1)$-cell, with boundary points x_0', x_1', x_2' and interior points y_0', y_1', y_2', and $x_0' = x_0$. Then the λ_{m-1}-eigenvalue equation says

(3.2.8) $$(4-\lambda_{m-1})u(x_0) = u(x_1) + u(x_2) + u\left(x_1'\right) + u\left(x_2'\right)$$

and the λ_m-eigenvalue equation says

(3.2.9) $$(4-\lambda_m)u(x_0) = u(y_1) + u(y_2) + u\left(y_1'\right) + u\left(y_2'\right).$$

Now each term on the right side of (3.2.9) is given by (3.2.7), and the sum is

$$(3.2.10) \qquad \frac{(6-\lambda_m)\left(u(x_1)+u(x_2)+u\left(x_1'\right)+u\left(x_2'\right)\right)+4(4-\lambda_m)u(x_0)}{(2-\lambda_m)(5-\lambda_m)}.$$

When we substitute (3.2.8) in (3.2.10) we obtain

$$(3.2.11) \qquad \left(\frac{(6-\lambda_m)(4-\lambda_{m-1})+4(4-\lambda_m)}{(2-\lambda_m)(6-\lambda_m)}\right)u(x_0).$$

So (3.2.9) will be valid provided

$$(3.2.12) \qquad 4-\lambda_m = \frac{(6-\lambda_m)(4-\lambda_{m-1})+4(4-\lambda_m)}{(2-\lambda_m)(5-\lambda_m)}$$

(this condition is also clearly necessary, unless $u(x_0)=0$ for all $x_0 \in V_{m-1} \setminus V_0$).

It is possible to simplify (3.2.12), but first we must record one more forbidden eigenvalue, $\lambda_m \neq 6$ (if $\lambda_m = 6$ then (3.2.12) is true, regardless of the value of λ_{m-1}). Then (3.2.12) becomes

$$((2-\lambda_m)(5-\lambda_m)-4)(4-\lambda_m) = (6-\lambda_m)(4-\lambda_{m-1}),$$

and since $(2-\lambda_m)(5-\lambda_m)-4 = (6-\lambda_m)(1-\lambda_m)$, this becomes

$$(6-\lambda_m)(1-\lambda_m)(4-\lambda_m) = (6-\lambda_m)(4-\lambda_{m-1}).$$

Since $\lambda_m \neq 6$ we may cancel the factor $6-\lambda_m$ to obtain $(1-\lambda_m)(4-\lambda_m) = 4-\lambda_{m-1}$, and finally

$$(3.2.13) \qquad \lambda_{m-1} = \lambda_m(5-\lambda_m).$$

This is the analog of (3.1.8). We can also solve

$$(3.2.14) \qquad \lambda_m = \frac{5+\varepsilon_m\sqrt{25-4\lambda_{m-1}}}{2} \quad \text{for } \varepsilon_m = \pm 1.$$

LEMMA 3.2.1 *Suppose $\lambda_m \neq 2, 5$, or 6, and λ_{m-1} is given by (3.2.13). (a) If $u|_{V_{m-1}}$ is a λ_{m-1}-eigenfunction of Δ_{m-1} and is extended to V_m by (3.2.7), then it is a λ_m-eigenfunction of Δ_m. (b) Conversely, if $u|_{V_m}$ is a λ_m-eigenfunction of Δ_m, then $u|_{V_{m-1}}$ is a λ_{m-1}-eigenfunction of Δ_{m-1}.*

Proof: Part (a) follows from the above discussion. For part (b), since the λ_m-eigenvalue equation holds, we have (3.2.7). Then the reasoning from (3.2.8) to (3.2.9) may be reversed if (3.2.12) holds, so (3.2.8) follows from (3.2.9) and (3.2.12). $\qquad \square$

Of course, if $m = 1$ there is no equation on V_0, so (a) above is valid without any assumption on $u|_{V_0}$.

Next we want to take the limit as $m \to \infty$. We assume that we have an infinite sequence $\{\lambda_m\}_{m \geq m_0}$ related by (3.1.13) or (3.1.14), with all but a finite number of $\varepsilon_m = -1$. Then we may define

$$(3.2.15) \qquad \lambda = \frac{3}{2}\lim_{m\to\infty} 5^m\lambda_m.$$

It is easy to see that the limit exists since

(3.2.16) $$\frac{5-\sqrt{25-4x}}{2} = \frac{1}{5}x + O\left(x^2\right) \quad \text{as } x \to 0.$$

Now suppose we start with a λ_{m_0}-eigenfunction u of Δ_{m_0} on V_{m_0}, and extend u to V_* by successively using (3.2.7), assuming that none of the λ_m is a forbidden eigenvalue (2, 5, 6). (If $m_0 = 1$ we can choose any values of $u\big|_{V_0}$ and extend to V_1 by (3.2.7).) Since (3.2.16) implies $\lambda_m = O(\frac{1}{5^m})$ as $m \to \infty$, it is easy to see that u is uniformly continuous on V_* and so extends to a continuous function on K. Moreover, it satisfies the λ-eigenvalue equation for Δ. This is the generic case of the spectral decimation recipe.

It is clear that this recipe constructs many eigenfunctions of Δ, but does it construct them all? It would be nice to have a direct argument to show that any eigenfunction of Δ must restrict to an eigenfuncton of Δ_m for m large enough, but it seems unlikely that such an argument exists (for I, we were able to use known properties of exponentials, sines, and cosines). We will eventually answer this question in the next section.

First we consider "small" eigenvalues. Suppose we start with $|\lambda_1| < 2$ and choose all $\varepsilon_m = -1$. (Here we allow complex eigenvalues and eigenvectors.) It is easy to see that $|\lambda_m| < 2$ for all m, so we never encounter a forbidden eigenvalue. Therefore, the recipe produces a three-dimensional space of eigenfunctions, with arbitrary boundary values. But the dimension of the space of eigenfunctions for a fixed eigenvalue is at most three, unless the eigenvalue is a Dirichlet eigenvalue (a four-dimensional space must contain a nonzero function vanishing on V_0). In the next section we will see that these eigenvalues are never Dirichlet eigenvalues. So we have constructed all eigenfunctions for these eigenvalues.

What exactly is this set of eigenvalues? It is difficult to give an exact description, but we can give a qualitative description. Define

(3.2.17) $$\varphi_\pm(x) = \frac{5 \pm \sqrt{25 - 4x}}{2}$$

and

(3.2.18) $$\Phi(z) = \frac{3}{2} \lim_{m \to \infty} 5^m \varphi_-^{(m)}(z) \quad (m\text{-fold iterate}).$$

Using (3.2.16) it is not difficult to show that Φ is an analytic function in the complement of $[\frac{25}{4}, \infty)$ with $\Phi(0) = 0$, $\Phi'(0) \neq 0$. Since the set of small eigenvalues is $\Phi(B_2)$, for B_2 the open ball of radius 2 about 0, we know that it forms a bounded open neighborhood of 0. In particular, it contains B_ε for some $\varepsilon > 0$.

Now we consider the set of eigenvalues obtained by starting with λ_{m_0} satisfying $|\lambda_{m_0}| < 2$ and choosing $\varepsilon_m = -1$ for all $m > m_0$. It is clear that we produce the set $5^{m_0}\Phi(B_2)$, which contains $B_{5^{m_0}\varepsilon}$. There are two obstacles to repeating the previous arguments. First, in order to get started, we need a λ_{m_0}-eigenfunction of Δ_{m_0} with prescribed boundary values. By linear algebra this is possible as long as λ_{m_0} is not a Dirichlet eigenvalue of Δ_{m_0}. In the next section we will see that the smallest Dirichlet eigenvalue of Δ_{m_0} is 2 or 5, so this is no problem after all. The other obstacle is that λ might be a Dirichlet eigenvalue of Δ. This is a legitimate problem, so we postpone it to the next section. We can now summarize what we have found.

THEOREM 3.2.2 *Suppose λ is not a Dirichlet eigenvalue of Δ. Then there exists m_0 (depending explicitly on $\log|\lambda|$) such that for any λ-eigenfunction u, $u\big|_{V_{m_0}}$ is an eigenfunction of Δ_{m_0}, and u is constructed by the spectral decimation recipe. The space of λ-eigenfunctions is exactly three-dimensional, and an eigenfunction is uniquely determined by its boundary values.*

It is worth pondering the similarities and differences between the functions $x(4-x)$ and $x(5-x)$ that govern the dynamics of the mapping $\lambda_m \to \lambda_{m-1}$ in the two cases, I and SG. If you are looking for strange numerology you could make the observation that this is, once again, just a matter of replacing 4 by 5 (we saw this in the pointwise definition of the Laplacian). Perhaps this is just a coincidence, perhaps not. But there is a big difference between these two quadratic polynomials. The Julia set of $x(4-x)$ is the interval $[0,4]$, while the Julia set of $x(5-x)$ is a Cantor set. This will have interesting implications for the spectrum of Δ on SG.

EXERCISES

3.2.1. Show that the matrix

$$\begin{pmatrix} 4 & -1 & -1 \\ -1 & 4 & -1 \\ -1 & -1 & 4 \end{pmatrix}$$

has eigenvalues 2 and 5 (multiplicity 2). Relate this fact to the solution of (3.2.3).

3.2.2. If $\lambda_m = 2$ use (3.2.4) to conclude that $u(x_0) + u(x_1) + u(x_2) = 0$. If $\lambda_m = 5$ use (3.2.6) to conclude that $u(x_0) = u(x_1) = u(x_2)$.

3.2.3. If λ is a negative real, show that all λ_m are negative reals.

3.2.4. Prove (3.2.16), and use it to show that the limit in (3.2.15) exists under the condition that all but a finite number of $\varepsilon_m = -1$. Also show that the limit in (3.2.15) does not exist if there are an infinite number of $\varepsilon_m = 1$.

3.2.5. Show that the spectral decimation recipe always produces a uniformly continuous function on V_*, and $-\Delta u = \lambda u$.

3.2.6. Show that (3.2.18) defines an analytic function in the complement of $[\frac{25}{4}, \infty)$ with $\Phi(0) = 0$, $\Phi'(0) \neq 0$.

3.2.7. Give an explicit estimate for m_0 in Theorem 3.2.2.

3.2.8. If $-\Delta u = \lambda u$, show that

$$-\Delta(u \circ F_w) = \frac{\lambda}{5^{|w|}} u \circ F_w.$$

3.2.9.* Show that if λ is negative and $-\Delta u = \lambda u$, then u is the minimizer of $\mathcal{E}(v, v) - \lambda \int_K v^2 d\mu$ among all functions v with the same boundary values. Also show that a minimizer exists.

3.3 EIGENVALUES AND MULTIPLICITIES

In this section we find all Dirichlet eigenvalues and eigenfunctions and their multiplicities. Not surprisingly, this will involve using the forbidden eigenvalues. The first problem we consider is to describe the spectrum of Δ_m. We will consider two kinds of eigenvalues, *initial* and *continued*. The continued eigenvalues will be those that arise from eigenvalues of Δ_{m-1} by the spectral decimation formula. Those that remain, the initial eigenvalues, must be some of the forbidden eigenvalues by Lemma 3.2.1.

So we begin with Δ_1. We find by inspection a one–dimensional Dirichlet eigenspace for $\lambda_1 = 2$ and a two-dimensional Dirichlet eigenspace for $\lambda_1 = 5$, shown in Figure 3.3.1. Since $\#(V_1 \setminus V_0) = 3$, this is the complete spectrum. Note that these eigenspaces transform under the symmetry group D_3 according to the trivial and two-dimensional representations. Incidentally, if we drop the Dirichlet condition we find two more eigenfunctions with eigenvalue 2, and just one more eigenfunction with eigenvalue 5, because (3.2.6) implies that all the boundary values are equal. See Figure 3.3.2. We also note that the forbidden eigenvalue 6 occurs with multiplicity 3 as shown in Figure 3.3.3, but it is not a Dirichlet eigenvalue.

Next we consider Δ_2. First we observe that there are no Dirichlet eigenfunctions corresponding to $\lambda_2 = 2$. This follows from (3.2.4), which says that the sum of the

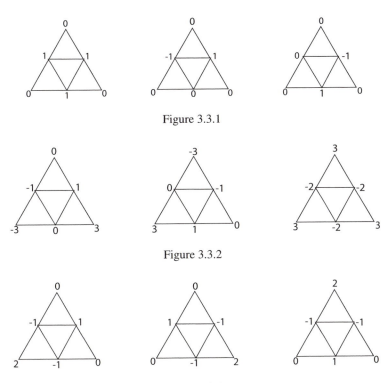

Figure 3.3.1

Figure 3.3.2

Figure 3.3.3

boundary values on each of the three 1-cells $F_i V_1$ must vanish. So if u vanishes on V_0, we have

$$u(x_0) + u(x_1) = u(x_1) + u(x_2) = u(x_0) + u(x_2) = 0,$$

and so u vanishes on V_1. That means u restricted to each 1-cell $F_i V_1$ must be a multiple of the Dirichlet 2-eigenvalue in Figure 3.3.1. It is easy to see that it is impossible to make the 2-eigenvalue equation hold at the points in $V_1 \setminus V_0$ unless all the multiples are 0, so $u \equiv 0$. This argument may be used inductively to prove that there are no $\lambda_m = 2$ Dirichlet eigenvalues of Δ_m for any $m > 1$. We leave the details to the exercises.

Moving on to the $\lambda_2 = 5$ case, we look for ways to duplicate the construction in Figure 3.3.1 on a smaller level. We note that (3.2.6) implies that u must vanish on V_1 for a Dirichlet eigenfunction, so this is the only possible approach. Note that we have a two-dimensional space of possibilities for the restriction of u to each 1-cell $F_i V_1$, and within this six-dimensional space there are three linear constraints expressing the 5-eigenvalue equation at each point in $V_1 \setminus V_0$. So we expect to find a three-dimensional space of Dirichlet eigenfunctions, and indeed we find them as shown in Figure 3.3.4 (to make these figures more legible we omit all labels of values 0). One nice metaphor for this construction is a "chain of batteries." Each battery has a positive and negative terminal, and it must be aligned with neighboring batteries positive to negative, or a charged terminal may abut a boundary point. If we take the sum of the three eigenfunctions in Figure 3.3.4, we obtain a circuit of three batteries around the central empty upside-down triangle, as shown

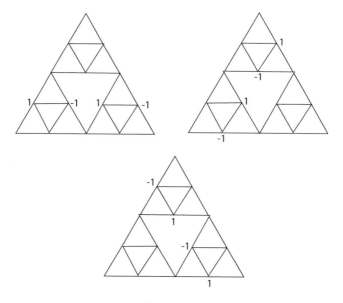

Figure 3.3.4

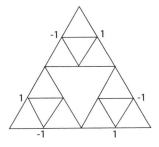

Figure 3.3.5

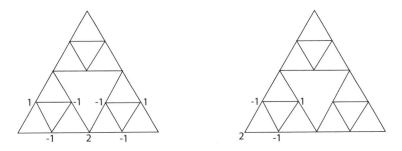

Figure 3.3.6

in Figure 3.3.5. In order to generalize this construction to higher values of m, we will take as our basis the eigenfunction in Figure 3.3.5 together with the first two in Figure 3.3.4, even though this breaks the symmetry.

Next we look for the forbidden eigenvalue $\lambda_2 = 6$. It is convenient to drop the Dirichlet condition and first look for all solutions of the 6-eigenvalue equation. Clearly the restriction to each 1-cell $F_i V_1$ must be a 6-eigenfunction on V_1, and these are described in Figure 3.3.3. Since we must glue these together at the points in $V_1 \setminus V_0$, there is at most one eigenfunction with values prescribed at the points in V_1. Miraculously they all turn out to satisfy the 6-eigenvalue equation at the points of $V_1 \setminus V_0$. (An explanation of the miracle will come later.) So we obtain a six-dimensional space of eigenfunctions, with a basis indexed by points in V_1, with $u_j(x_k) = \delta_{jk}$ if $V_1 = \{x_1, \ldots, x_6\}$. We show two of these eigenfunctions in Figure 3.3.6, and the others are obtained by rotation. By inspection we see that there is a three-dimensional space of Dirichlet eigenfunctions indexed by the points in $V_1 \setminus V_0$. From $\lambda_1 = 2$ we obtain $\lambda_2 = \frac{5 \pm \sqrt{17}}{2}$ with multiplicities 1, and from $\lambda_1 = 5$ we obtain $\lambda_2 = \frac{5 \pm \sqrt{5}}{2}$ with multiplicities 2 (the multiplicity of $\lambda_1 = 5$).

Now for some arithmetic. We have found initial eigenvalues $\lambda_2 = 5$ and $\lambda_2 = 6$ with multiplicities 3 and 3. We also have continued eigenvalues obtained from $\lambda_1 = 2, 5$ by spectral decimation. From $\lambda_1 = 2$ we obtain $\lambda_2 = \frac{5 \pm \sqrt{17}}{2}$ with multiplicities 1, and from $\lambda_1 = 5$ we obtain $\lambda_2 = \frac{5 \pm \sqrt{5}}{2}$ with multiplicities 2 (the multiplicity of $\lambda_1 = 5$). The grand total is $3 + 3 + 1 + 1 + 2 + 2 = 12$, and $12 = \#(V_2 \setminus V_0)$, the dimension of the space for the Dirichlet Laplacian Δ_2. This verifies that we have found the complete Dirichlet spectrum of Δ_2. In increasing order it reads

$$\left(\frac{5-\sqrt{17}}{2}, \frac{5-\sqrt{5}}{2}, \frac{5+\sqrt{5}}{2}, \frac{5+\sqrt{17}}{2}, 5, 6\right)$$

with multiplicities $(1, 2, 2, 1, 3, 3)$. We also observe that Lemma 3.2.1 does imply that the entire spectrum of Δ_2 must arise in this way, so the dimension count is just a check. In the general case it will be more difficult to prove that we have correctly computed the multiplicities of the initial eigenvalue 5, so we will rely on a dimension count to complete the argument.

Now we are ready for the general case. We know that $\lambda_m = 2$ is not a Dirichlet eigenvalue, so $\lambda_m = 5$ and $\lambda_m = 6$ are the initial eigenvalues, and we need to compute their multiplicities $M_m(5)$ and $M_m(6)$. By repeating the analysis of the 6-eigenfunctions of Δ_2, we see that $M_m(6) = \#(V_{m-1} \setminus V_0) = \frac{3^m - 3}{2}$ (see Exercise 1.1.1) with an eigenfunction associated to each point x in $V_{m-1} \setminus V_0$, namely the first eigenfunction in Figure 3.3.6 reduced and rotated to fit the point. If $x = F_w q_i = F_{w'} q_{i'}$ for $|w| = |w'| = m - 1$, then the eigenfunction u is supported in $F_w V_1 \cup F_{w'} V_1$. This is the first example of a completely localized eigenfunction. We will discuss this phenomenon in detail in the next section.

Using the battery metaphor we may describe a large set of linearly independent 5-eigenfunctions. This will give us a lower bound for $M_5(m)$, and later the dimension count will show that this lower bound is the correct value. The idea is based on the 1-homology of the graph Γ_{m-1}. If you look at the graph you see a bunch of 1-cycles going around each of the empty upside-down triangles of different sizes. Clearly these cycles are not boundaries, and it is not hard to see that they are linearly independent. (If you delete the boundary points, these will generate the entire 1-homology group.) For each cycle we may put a string of batteries around it, the number of batteries depending on the size of the cycle. In this way we obtain 5-eigenfunctions associated to each cycle. We also want to consider two more, a string of batteries joining q_0 to q_1, and a string of batteries joining q_1 to q_2. It is possible to show that this yields a linearly independent set of 5-eigenfunctions, but we leave the details to the exercises. Note that we would not want to adjoin another string of batteries joining q_2 to q_0, for then we would not have a linearly independent set (the total sum, with appropriate signs, would be zero). The total number of cycles, counting down in size, is $1 + 3 + 3^2 + \cdots + 3^{m-2} = \frac{3^{m-1}-1}{2}$, so our lower bound is $M_5(m) \geq \frac{3^{m-1}+3}{2}$.

So that is the story for initial eigenvalues. What about continued eigenvalues? It is essentially the same as for $m = 2$, except there is one new twist. If $\lambda_{m-1} = 6$, then the two solutions (3.2.14) for λ_m are 2 and 3, and 2 is a forbidden eigenvalue. (On the other hand, we never find $\lambda_m = 5$ or 6 in (3.2.14), because that would make $\lambda_{m-1} = 0$ or -6, and all Dirichlet eigenvalues are positive.) So every eigenvalue λ_{m-1} bifurcates into two choices of λ_m, by the choice $\varepsilon_m = \pm 1$, except $\lambda_{m-1} = 6$, which just yields the single choice $\lambda_m = 3$ corresponding to $\varepsilon_m = 1$.

The total space of Dirichlet eigenfunctions for Δ_{m-1} has dimension $\frac{3^m-3}{2}$, and for Δ_m it is $\frac{3^{m+1}-3}{2}$ (see Exercise 1.1.1). Of the $\frac{3^m-3}{2}$ eigenfunctions of Δ_{m-1}, we know that $M_6(m-1) = \frac{3^{m-1}-3}{2}$ of them correspond to eigenvalue $\lambda_{m-1} = 6$, while the remaining $\frac{3^m - 3^{m-1}}{2} = 3^{m-1}$ of them correspond to other eigenvalues, leading to a space of continued eigenfunctions of dimension $\frac{3^{m-1}-3}{2} + 2 \cdot 3^{m-1}$. If we add to this $M_6(m) = \frac{3^m - 3}{2}$ and the lower bound $\frac{3^{m-1}+3}{2}$ for $M_5(m)$, we obtain

$$\frac{3^{m-1}-3}{2} + \frac{4 \cdot 3^{m-1}}{2} + \frac{3^m - 3}{2} + \frac{3^{m-1}+3}{2} = \frac{9 \cdot 3^{m-1} - 3}{2} = \frac{3^{m+1}-3}{2}$$

as desired. So by induction we have the complete spectrum.

Having found the Dirichlet eigenvalues and eigenfunctions for Δ_m, it follows by routine limiting arguments that the Dirichlet spectrum of Δ is obtained in the limit as $m \to \infty$. Note that each Dirichlet eigenvalue λ of Δ enters as λ_m in the Dirichlet spectrum of Δ_m with the correct multiplicity. We leave the details to the exercises.

One more small technical point remains. We have shown how to construct all Dirichlet eigenfunctions corresponding to a Dirichlet eigenvalue λ, and the spectral decimation recipe is still valid, starting with a generation of birth m_0 where $\lambda_{m_0} = 2, 5$, or 6. Can we say the same for all eigenfunctions corresponding to that eigenvalue λ? The proof of Theorem 3.2.2 does not answer this question. We can show that the answer is yes by a case-by-case analysis. If $\lambda_{m_0} = 6$, then there are three more eigenfunctions corresponding to the vertices in V_0 (see Figure 3.3.3 for $m_0 = 1$ and the right side of Figure 3.3.6 for $m_0 = 2$). This means that spectral decimation constructs λ-eigenfunctions with arbitrary boundary values. Therefore, an arbitrary λ-eigenfunction is a sum of one of these and a Dirichlet eigenfunction, and hence is constructed by spectral decimation. The case $\lambda_1 = 2$ or 5 is more delicate. Note that the eigenfunctions shown in Figures 3.3.1 and 3.3.2 together span a three-dimensional space in each case, but we cannot prescribe boundary values freely. In the case $\lambda_1 = 2$ we have the constraint $u(q_0) + u(q_1) + u(q_2) = 0$, and in the case $\lambda_1 = 5$ we have the constraint $u(q_0) = u(q_1) = u(q_2)$. These constraints were derived from the spectral decimation recipe. How do we know they hold for all λ-eigenfunctions? A direct proof would be difficult, so we argue indirectly. Each eigenfunction u is made up by gluing together the functions $u \circ F_i$, which are $\frac{\lambda}{5}$-eigenfunctions. Since $\frac{\lambda}{5}$ is not a Dirichlet eigenvalue, the functions $u \circ F_i$ are determined by their boundary values $u(F_i q_j)$ by Theorem 3.2.2. So there are six values to be prescribed at the points in V_1. From our gluing theorem we know that we get an eigenfunction if and only if the matching conditions for normal derivatives hold at the points in $V_1 \setminus V_0$. Note that this condition is independent of the spectral decimation recipe. We have not explained how to compute these normal derivatives yet. A priori they might all vanish, but we can show that this doesn't happen. Indeed, if all the normal derivatives vanished, then the matching conditions would be automatic, so there would be a three-dimensional space of Dirichlet eigenfunctions. But we know this is false. Next, by using symmetry considerations, we can show that the matching conditions for normal derivatives at a vertex in $V_1 \setminus V_0$ is just a multiple of the 5-eigenvalue equation for Δ_1 at that point. We leave the details to the exercises. So every λ-eigenfunction on K restricts to a λ_1-eigenfunction on V_1, and the spectral decimation recipe works. Similar reasoning works in the case $\lambda_{m_0} = 5$ for any m_0.

In summary, the first conclusion of Theorem 3.2.2 remains true if λ is a Dirichlet eigenvalue; the only differences are that the space of all eigenfunctions need not have dimension 3, and boundary values cannot be freely prescribed.

We will examine the Dirichlet spectrum more closely in later sections. Here we will just make a few observations. We may classify eigenvalues into three series, which we call the 2-series, 5-series, and 6-series, depending on the value of λ_{m_0}. The eigenvalues in the 2-series all have multiplicity 1, while the eigenvalues in the other series all exhibit higher multiplicity. Also, if λ is an eigenvalue in the 5-series or 6-series, then $5^m \lambda$ is also an eigenvalue, corresponding to a generation of birth $m_0 + m$, with the same choice of ε's (suitably reindexed). The lowest eigenvalue, or *bottom of the spectrum*, belongs to the 2-series with all $\varepsilon_m = -1$. The corresponding eigenfunction, called the *ground state*, is positive (see the exercises).

Next we give a brief discussion of the Neumann spectrum of Δ. As indicated in Section 3.1, we want to impose a Neumann condition on the graph Γ_m by imagining that it is embedded in a larger graph by reflecting in each boundary vertex and imposing the λ_m-eigenvalue equation on the even extension of u. This just means that we impose the equation

$$(3.3.1) \qquad (4 - \lambda_m) u (q_0) = 2u \left(F_0^m q_1 \right) + 2u \left(F_0^m q_2 \right)$$

at q_0, and so on. Then the Neumann λ_m-eigenvalue equations consist of exactly $\#V_m$ equations in $\#V_m$ unknowns. It is even convenient to allow $m = 0$, in which case there are three equations (3.3.1) and no others. In particular, on V_0 we find eigenvalues $\lambda_0 = 0$ corresponding to the constant function, and $\lambda_0 = 6$ corresponding to the two-dimensional space of functions satisfying $u(q_0) + u(q_1) + u(q_2) = 0$.

When $m = 1$, we find a three-dimensional space with $\lambda_1 = 6$, namely the ones described in Figure 3.3.3, since we can check that these satisfy condition (3.3.1). This is the secret behind the ease with which we glued small copies of these eigenvalues together: They have normal derivatives equal to zero, so the matching condition for normal derivatives is automatic. This is the only initial eigenvalue. There are unique ways to continue $\lambda_0 = 0$ and $\lambda_0 = 6$, namely $\lambda_1 = 0$ (5 is forbidden) and $\lambda_1 = 3$, so this completes the spectrum of Δ_1.

The general case is very similar to the Dirichlet spectrum, with only a few changes:

(1) The constant function is a Neumann eigenfunction with all $\lambda_m = 0$ and $\lambda = 0$.

(2) There is no 2-series.

(3) The 6-series has multiplicity increased by 3, namely the eigenfunctions associated to the boundary points.

(4) The 5-series has multiplicity reduced by 2, since we retain all eigenfunctions associated with loops, but discard the two extra ones that chain from one boundary point to another (these do not satisfy (3.3.1)). In particular, the 5-series begins with $m_0 = 2$. We leave it to the exercises to verify that the dimensions add up correctly.

We still have to justify (3.3.1). If it holds for all m, then since $\lambda_m = O(\frac{1}{5^m})$ for large m we can use it to obtain $\partial_n u(q_i) = 0$, the usual expression of the Neumann boundary condition. Conversely, $\partial_n u(q_i) = 0$ allows us to glue the function to its even reflection, and that leads to (3.3.1).

EXERCISES

3.3.1. Show that there are no Dirichlet eigenvalues with $\lambda_{m_0} = 2$ for $m_0 > 1$. Show that there are no Neumann eigenvalues with $\lambda_{m_0} = 2$ for any m_0.

3.3.2. Prove the linear independence of the $\frac{3^{m-1}+3}{2}$ Dirichlet 5-eigenfunctions for Δ_m constructed above.

3.3.3.* Prove that the Dirichlet (resp., Neumann) spectrum of Δ is the limit (in the sense of (3.2.15)) of the Dirichlet (resp., Neumann) spectra of Δ_m.

3.3.4.* Show directly that the matching conditions for normal derivatives for u at a point in $V_1 \setminus V_0$ is just a multiple of the λ_1-eigenvalue equation at that point if $\lambda_1 = 2$ or 5.

3.3.5. Show that the ground state Dirichlet eigenfunction is positive on $K \setminus V_0$.

3.3.6. Verify that the sum of the multiplicities of the Neumann eigenvalues of Δ_m listed above adds to $\#V_m = \frac{3^{m+1}+3}{2}$.

3.3.7. Show that every Dirichlet eigenfunction in the 2-series is actually constant along the upside-down triangle joining the points in $V_1 \setminus V_0$. (Hint: Use (3.2.7) and induction.)

3.3.8.* Show that every Dirichlet eigenfunction in the 2-series is invariant under the discontinuous map that reflects each cell $F_i K$ about the axis through q_i. Then show by induction that it is invariant under an infinite sequence of discontinuous maps that reflect the $3 \cdot 2^{m-1}$ m-cells that line the upside-down triangle in Exercise 3.3.7. Use this to give a different proof of Exercise 3.3.7.

3.3.10.* Let $\tilde{\lambda}_1 < \tilde{\lambda}_2 < \cdots$ denote the distinct Dirichlet eigenvalues of Δ. Show that they have the following "octave" structure: Of the eigenvalues $\tilde{\lambda}_{7k+1}, \tilde{\lambda}_{7k+2}, \ldots, \tilde{\lambda}_{7k+7}$, the first four have generation of birth $m_0 = 1$, with $\tilde{\lambda}_{7k+1}$ and $\tilde{\lambda}_{7k+4}$ belonging to the 2-series, while $\tilde{\lambda}_{7k+2}$ and $\tilde{\lambda}_{7k+3}$ belong to the 5-series; $\tilde{\lambda}_{7k+5}$ and $\tilde{\lambda}_{7k+7}$ also belong to the 5-series but with $m_0 > 1$, and $\tilde{\lambda}_{7k+6}$ belongs to the 6-series with generation of birth $m_0 - 1$.

3.3.11. Show that the sequence $\{\frac{3}{2}5^m \lambda_m\}$ is increasing to λ for every Dirichlet eigenvalue.

3.4 LOCALIZED EIGENFUNCTIONS

We observed in the last section that there exist localized eigenfunctions on SG, meaning eigenfunctions with small support. This is true for the entire basis of eigenfunctions in the 6-series with large value of m_0, the support being the union of two adjacent m_0-cells. It is also true for some of the basis functions in the 5-series for large m_0, namely those associated to small loops. Since the existence of localized eigenfunctions is unprecedented in all of smooth mathematics, it is worth while trying to understand what makes it possible. We observe from the start that localized eigenfunctions are associated with eigenvalues of high multiplicities, and also eigenfunctions that are simultaneously Dirichlet and Neumann. (By the way, on I the nonzero eigenvalues are all both Dirichlet and Neumann, but there are no joint Dirichlet and Neumann eigenfunctions, and of course no localized eigenfunctions.)

Both I and SG are self-similar and highly symmetric. One aspect of the sine and cosine eigenfunctions on I that we did not emphasize in Section 3.1 is that they are all locally equivalent. For example, $\sin \pi k x$ is obtained from $\sin \pi x$ by composing with the mappings $x \to kx - j$ on the interval $[\frac{j}{k}, \frac{j+1}{k}]$, multiplying by $(-1)^j$ and gluing. Note that this exploits the self-similarity of I with respect to more mappings than just the dyadic similarities. It is not true on SG that all eigenfunctions may be constructed in this fashion, but it is true for some of them. The key idea is that if we start with a λ-eigenfunction u that is both a Dirichlet and Neumann eigenfunction, then $u \circ F_w^{-1}$ on $F_w K$ satisfies the $5^{|w|}\lambda$-eigenvalue equation on $F_w K$ and may be glued to the zero function outside $F_w K$ to yield a joint Dirichlet–Neumann (D–N for short) eigenfunction on K with eigenvalue $5^{|w|}\lambda$ and with support in $F_w K$. If we fix m and look at all words with $|w| = m$, we see that there are at least 3^m localized D–N eigenfunctions (supported in an m-cell) associated to the eigenvalue $5^m \lambda$. So we have simultaneously created localized eigenfunctions and high multiplicities. In addition, we have shown that 5 is a spectral multiplication factor, at least for a portion of the spectrum.

For example, if we start with $m_0 = 2$ and $\lambda_2 = 6$, the three Dirichlet eigenfunctions of Δ_2 (illustrated on the left in Figure 3.3.6) are also Neumann eigenfunctions, so all eigenfunctions of Δ generated by spectral decimation from these (there are infinitely many, since we have the choices of ε_m for $m \geq 4$) are D–N eigenfunctions. The above procedure produces eigenvalues with $m_0 = 2 + m$ and $\lambda_{2+m} = 6$. Note, however, that it does not generate all the localized eigenfunctions with the given eigenvalue. We had one Dirichlet eigenfunction for each point x in V_{m+1}, supported in the two $(m + 1)$-cells touching at x. If $x \notin V_m$, then the two cells lie in a single m-cell, and these eigenfunctions are the ones constructed above. But if $x \in V_m$, then the two $(m + 1)$-cells lie in distinct m-cells, and the eigenfunction is not of the above type.

Similarly, we see that there are localized eigenfunctions in the 5-series, since there is a D–N eigenfunction with $m_0 = 2$ and $\lambda_2 = 5$, namely the one shown in Figure 3.3.5. By localizing this eigenfunction we get all the eigenfunctions associated with loops of minimal size (for the given value of m_0). There are also localized eigenfunctions associated with larger loops that are not simply localizations of this single eigenfunction.

So far we have seen that the existence of a D–N eigenfunction gives us a seed for constructing some of the localized eigenfunctions we know about. But why should there exist D–N eigenfunctions? One answer is that high multiplicities require it. Specifically, if λ is a Dirichlet eigenvalue of multiplicity at least 4, then there must be a D–N eigenfunction in the eigenspace. This is just linear algebra: The Neumann conditions are just three homogeneous linear equations looking for a nontrivial solution. So the existence of one eigenvalue with multiplicity at least 4 implies not only localized eigenfunctions but also arbitrarily large multiplicities higher up in the spectrum.

Although this is an interesting observation, it does not appear to be very useful, because it is difficult to find an a priori argument for multiplicities above 2. But there is a reason for the existence of D–N eigenfunctions based solely on the D_3 symmetry of SG. The dihedral group has a nontrivial one-dimensional representation, called the alternating representation. This is just a funny way of saying that there are nontrivial functions that are invariant under rotations and skew-symmetric under reflections. We say these functions *transform according to the alternating representation* of D_3. The function in Figure 3.3.5 is one example. Now a general principle in group theory says that there must be infinitely many eigenfunctions of Δ within this class of functions. But the skew-symmetry under reflections implies that both the function and its normal derivative must vanish at boundary points. So all the eigenfunctions of this type are D–N. This argument applies to all fractal Laplacians that have dihedral symmetry of any order, even without spectral decimation. We will return to this in Chapter 4.

How many localized eigenfunctions are there? We don't mean this question literally, but rather: What is the proportion of linearly independent localized eigenfunctions among all Dirichlet (or Neumann) eigenfunctions with eigenvalues going up to a fixed value? Because we also have to quantify the degree of localization, the answer is complicated, so we look at a simpler question: How many eigenfunctions are D–N? The surprising answer is: almost all!

How do we figure out what a bottom part of the Dirichlet spectrum looks like? One simple way is to look at the spectrum of Δ_m. It consists of $\frac{3^{m+1}-3}{2}$ eigenvalues, counting multiplicity, and of these $\frac{3^m-3}{2}$ correspond to $\lambda_m = 6$. If we extend these eigenvalues by choosing $\varepsilon_{m'} = -1$ for all $m' > m$ (except $\varepsilon_{m+1} = 1$ if $\lambda_m = 6$), then we will obtain the smallest continued eigenvalues, and in fact eigenvalues smaller than the initial eigenvalues for any $m_0 > m$. (Remember that initial eigenvalues for $m_0 \geq 2$ are either 5 or 6, and $\varphi_\pm(x) = \frac{5 \pm \sqrt{25-4x}}{2} < 5$ for $x > 0$.) Therefore, we obtain the lowest $\frac{3^{m+1}-3}{2}$ eigenvalues in the spectrum of Δ in the limit. We are not concerned here with the quantitative values of these eigenvalues, but rather with the qualitative features of the corresponding eigenfunctions, which are identical to the qualitative features of the eigenfunctions of Δ_m.

So the proportion of D–N eigenfunctions in the bottom part of the Dirichlet spectrum consisting of the first $\frac{3^{m+1}-3}{2}$ eigenvalues is identical to the proportion of D–N eigenfunctions in the Dirichlet spectrum of Δ_m. This is a question we can answer rather exactly.

Let a_m denote the total number of D–N eigenfunctions in the Dirichlet spectrum of Δ_m. For the initial eigenvalue $\lambda_m = 6$, all $\frac{3^m-3}{2}$ eigenfunctions are D–N. For the initial eigenvalue $\lambda_m = 5$, all those associated to loops are D–N, so the number is $\frac{3^{m-1}-1}{2}$. For the continued eigenvalues, we note that there are $a_{m-1} - \frac{3^{m-1}-3}{2}$ D–N eigenfunctions of Δ_{m-1} with $\lambda_{m-1} \neq 6$, and these bifurcate to produce $2a_{m-1} - 3^{m-1} + 3$ D–N eigenfunctions of Δ_m. The remaining $\frac{3^{m-1}-3}{2}$ D–N eigenfunctions of Δ_{m-1} with $\lambda_{m-1} = 6$ do not bifurcate, and so only produce $\frac{3^{m-1}-3}{2}$ D–N eigenfunctions of Δ_m. Adding everything up yields

$$
\begin{aligned}
a_m &= \frac{3^m - 3}{2} + \frac{3^{m-1} - 1}{2} + 2a_{m-1} - 3^{m-1} + 3 + \frac{3^{m-1} - 3}{2} \\
&= \frac{3^m - 1}{2} + 2a_{m-1}.
\end{aligned}
$$

(3.4.1)

If we let $b_m = \frac{3^{m+1}-3}{2} - a_m$ denote the number of remaining eigenfunctions, then (3.4.1) implies

$$
\begin{aligned}
b_m &= 3^m - 1 - 2a_{m-1} \\
&= 3^m - 1 - 2\left(b_{m-1} - \frac{3^m - 3}{2}\right) \\
&= 2b_{m-1} + 2,
\end{aligned}
$$

hence

(3.4.2) $$b_m = c2^m - 2$$

and $c = \frac{5}{2}$ because $b_1 = 3$. Clearly, the proportion $b_m/\frac{3^{m+1}-3}{2}$ of non-D–N eigenfunctions goes to zero at an exponential rate, so the proportion of D–N eigenfunctions goes to one.

This counting perspective also shows that the high multiplicities may be truly huge. If we look at the largest eigenvalue among the first $\frac{3^{m+1}-3}{2}$ Dirichlet eigenvalues ($m_0 = m$ and $\lambda_m = 6$), it has multiplicity $\frac{3^m-3}{2}$, so it accounts for almost $\frac{1}{3}$ of all these eigenvalues. We will return to this in the next section.

The existence of localized eigenfunctions has some strange consequences. Consider the heat equation

(3.4.3) $$\frac{\partial}{\partial t}u(x,t) = \Delta u(x,t),$$

where u is now a function on a subset of $SG \times \mathbb{R}$. This has solutions

(3.4.4) $$u(x,t) = e^{-\lambda t}u_\lambda(x),$$

where u_λ is a λ-eigenfunction. If u_λ is localized, then all the heat stays in supp u_λ and never leaks out to the rest of the world. This would indeed make a cozy cell, except that the temperature must be negative somewhere in the cell and it goes to zero in the long run! Similarly, the wave equation

(3.4.5) $$\frac{\partial^2}{\partial t^2}u(x,t) = \Delta u(x,t)$$

has solutions

(3.4.6) $$u(x,t) = \left(a\cos\sqrt{\lambda}t + b\sin\sqrt{\lambda}t\right)u_\lambda(x).$$

For u_λ localized this is a mode of vibration that only involves the cells in supp u_λ. Living in SG, you might not hear your noisy neighbors!

In smooth analysis, localized eigenfunctions cannot exist because eigenfunctions are analytic functions, and nonzero analytic functions cannot vanish on open sets (this assumes that the underlying manifold and Riemannian metric are analytic). So on SG, something must break down in the above reasoning. Actually there is a theory of analytic functions, but eigenfunctions are not necessarily globally analytic, although they are locally analytic.

EXERCISES

3.4.1.* Show that a priori there must be Dirichlet eigenvalues of multiplicity at least 2 because D_3 has an irreducible representation of dimension 2.

3.4.2. If γ_m denotes the largest Dirichlet eigenvalue among the first $\frac{3^{m+1}-3}{2}$ eigenvalues, show that $\gamma_m = c5^m$.

3.4.3. Find a formula for the number of distinct Dirichlet eigenvalues of Δ_m.

3.4.4.* Show that there are Dirichlet eigenfunctions in the 6-series that transform according to the alternating representation of D_3 with $m_0 \geq 3$, but not $m_0 = 2$.

3.4.5.* Show that there are no Dirichlet eigenfunctions in the 5-series that are invariant under D_3.

3.5 SPECTRAL ASYMPTOTICS

We have seen in the last section that the bottom $\frac{3^{m+1}-3}{2}$ eigenvalues of the Dirichlet spectrum of SG are generated from the spectrum of Δ_m. The largest of these eigenvalues (if $m \geq 2$) has $m_0 = m$, $\lambda_m = 6$, $\varepsilon_{m+1} = 1$, and $\varepsilon_{m'} = -1$ for $m' > m+1$. The eigenvalue $\lambda = \frac{3}{2} \lim_{m' \to \infty} 5^{m'} \lambda_{m'}$ is thus equal to a certain constant times 5^m. If we define the Dirichlet eigenvalue counting function

(3.5.1) $$\rho(x) = \#\{\lambda \in \Lambda_D : \lambda \leq x\},$$

where Λ_D denotes the Dirichlet spectrum (repeated according to multiplicity), then we have

(3.5.2) $$\rho\left(c_1 5^m\right) = \frac{3^{m+1} - 3}{2}$$

for the appropriate choice of c_1. This suggests an asymptotic growth rate

(3.5.3) $$\rho(x) \sim x^{\log 3 / \log 5}.$$

In fact the discussion shows that we expect more or less identical asymptotic behavior for the Neumann eigenvalue counting functions or the D–N eigenvalue counting function. We can ask, more precisely, for the behavior of the ratio

(3.5.4) $$\rho(x) / x^{\log 3 / \log 5}$$

as $x \to \infty$.

In analogy with the Weyl asymptotic law in smooth analysis, we might surmise that (3.5.4) converges to a limit, but we can immediately see that this is impossible because of the very high multiplicity of the eigenspace that gives rise to (3.5.2). Since this multiplicity is $\frac{3^m - 3}{2}$, the ratio (3.5.4) must jump apprecia-bly as x approaches $c_1 5^m$. So we need to look for a different type of behavior.

It is not difficult to show that (3.5.2) implies that the ratio (3.5.4) is bounded above and bounded away from zero. And, of course, there is a limit along the sequence $x = c_1 5^m$. In fact, what happens is that there is a limit along any sequence of the form $x = c 5^m$. What this means is that there is a periodic function $g(t)$ of period $\log 5$, such that

$$(3.5.5) \qquad \lim_{x \to \infty} \left(\frac{\rho(x)}{x^{\log 3 / \log 5}} - g(\log x) \right) = 0.$$

The function $g(t)$ is bounded above, bounded away from zero, and necessarily discontinuous at the value $\log c_1$.

Here we give a rough explanation of why this is so. An entirely different proof is sketched in the exercises. We will work with the D–N spectrum. Look at one particular eigenvalue x. It will have a generation of birth m_0 (with $\lambda_{m_0} = 5$ or 6) and also what we will call a *generation of fixation* m_1 defined to be the minimum value such that $\varepsilon_m = -1$ for all $m > m_1$. The location of x in the spectrum is determined by the location of λ_{m_1} in the spectrum of Δ_{m_1}. Note that $5x$ will also be an eigenvalue, with generation of birth $m_0 + 1$, generation of fixation $m_1 + 1$, and initial value and ε_m values in between being identical to those of x. We need to compare $\rho(x)$ and $\rho(5x)$. If we could show $\rho(5x) = 3\rho(x)$, then the ratio $\rho(x)/x^{\log 5 / \log 3}$ would be unchanged under $x \to 5x$. This is too much to expect, as (3.5.2) already shows, but we can hope that something very close to this is true.

As we have already observed, $\rho(x)$ is equal to the number of D–N eigenvalues of Δ_{m_1} with eigenvalue at most λ_{m_1}. And so $\rho(5x)$ is equal to the number of eigenvalues of Δ_{m_1+1} with eigenvalue at most the same λ_{m_1}. To a large extent, the distinct eigenvalues in the spectra of Δ_{m_1} and Δ_{m_1+1} that we are counting are identical, although their multiplicities are not the same. The exceptions are the eigenvalues in the spectrum of Δ_{m_1+1} that have generation of birth $m_0 = 2$ (this is the earliest for D–N eigenvalues). But these all have multiplicity at most 3, and there are at most 2^{m_1+1} of them, so they don't contribute anything to the ratio (2.5.4) in the limit. Every eigenvalue that enters into the count in the spectrum of Δ_{m_1} also enters into the count in the spectrum of Δ_{m_1+1}, but since its generation of birth is increased by 1, its multiplicity is essentially multiplied by 3. Again this is not precisely so, but the errors are small enough not to change the ratio in the limit. The exact statement is

$$(3.5.6) \qquad \rho(5x) = 3\rho(x) + O\left(m_1 2^{m_1}\right).$$

Since

$$(3.5.7) \qquad x^{\log 3 / \log 5} = O\left(3^{m_1}\right),$$

if m_1 is the generation of fixation of x, it follows easily that

$$(3.5.8) \qquad \frac{\rho(5^m x)}{(5^m x)^{\log 3 / \log 5}} = \frac{\rho(5^m x)}{3^m x^{\log 3 / \log 5}}$$

has a positive limit as $m \to \infty$.

This is not the entire story, because there are gaps. If x is as above and x' is the next D–N eigenvalue with generation of fixation also m_1, then there will be no D–N eigenvalues in between $5^m x$ and $5^m x'$ for any m. That means

(3.5.9) $$\rho(5^m t) = \rho\left(5^m x\right) \quad \text{for } x \le t < x',$$

and so

(3.5.10) $$\frac{\rho\left(5^m t\right)}{(5^m t)^{\log 3/\log 5}} = \frac{\rho\left(5^m x\right)}{(5^m x)^{\log 3/\log 5}} \left(\frac{x}{t}\right)^{\log 3/\log 5} \quad \text{for } x \le t < x'$$

also has a limit as $m \to \infty$. Note that this means the function

$$g(\log t) = ct^{-\log 3/\log 5}$$

is smooth and decreasing in this interval.

We also observe the same gap structure in the Dirichlet spectrum, where there are more distinct eigenvalues. All we have to do is take an x that is not in the 2-series, and let x' be the next distinct eigenvalue not in the 2-series. Then $5^m x$ and $5^m x'$ will be consecutive distinct eigenvalues for $m \ge 1$. These gaps are significant in comparison with the size of the eigenvalues,

(3.5.11) $$\frac{5^m x' - 5^m x}{5^m x} = \frac{x'}{x} \quad \text{independent of } m.$$

Another way of saying this is that they are visible on a logarithmic scale. We will discuss some of the interesting consequences of the gap structure of the spectrum in Chapter 5. Again, this is something unprecedented in smooth analysis. Note that on I the gap between $(\pi n)^2$ and $(\pi(n+1))^2$ is large, namely $2\pi^2 n$, but is small in comparison with the eigenvalue $(\pi n)^2$. This is typical for Laplacians on smooth manifolds.

What is the significance of the exponent $\log 3/\log 5$? In smooth analysis, the exponent is the ratio of the dimension of the space to the order of the operator. If we use the resistance metric, the dimension of SG is $\log 3/\log(5/3) = d$, as a ball of radius r will have measure $\approx r^d$ (Exercise 1.6.7). So the order of the Laplacian ought to be $\log 5/\log(5/3) = d + 1$, so $\log 3/\log 5 = d/(d+1)$. Note that this is consistent with the interval, where the Laplacian has order 2 and the dimension is 1.

EXERCISES

3.5.1. Show that the constant c_1 in (3.5.3) is given by $c_1 = \frac{3}{2} \lim_{m \to \infty} 5^{m+3} \varphi_-^{(m)}(3)$.

3.5.1. Show that (3.5.2) implies that the ratio (3.5.4) is bounded above and away from zero.

3.5.3. Prove (3.5.6).

3.5.4. Show that the function $g(t)$ has infinitely many jump discontinuities, but only a finite number with jump exceeding ε, for any fixed $\varepsilon > 0$.

3.5.5.* Let $\mathcal{R}(u) = \mathcal{E}(u)/\|u\|_2^2$ denote the Rayleigh quotient. Show that the standard minimax characterization of the Dirichlet $\{\lambda_j^D\}$ and Neumann $\{\lambda_j^N\}$ eigenvalues (repeated according to multiplicity) holds:

$$\lambda_j^D = \min_{L_j} \max \mathcal{R}(u),$$

where the minimum is over all j-dimensional subspaces L_j of $\mathrm{dom}_0\mathcal{E}$, and

$$\lambda_j^N = \min_{L_j'} \max \mathcal{R}(u),$$

where the minimum is taken over all j-dimensional subspaces L_j' of $\mathrm{dom}\,\mathcal{E}$.

3.5.6.* Show that $\lambda_j^N \leq \lambda_j^D$ and $\lambda_j^D \leq \lambda_{j+3}^N$. Conclude that

$$\rho^D(x) \leq \rho^N(x) \leq \rho^D(x)+3,$$

where ρ^D and ρ^N denote the eigenvalue counting functions for Dirichlet and Neumann eigenvalues, respectively. (Hint: Use Exercise 3.5.5.)

3.5.7.* Let $(\mathrm{dom}_0\mathcal{E})'$ denote the subspace (of codimension 3) of $\mathrm{dom}_0\mathcal{E}$ of functions vanishing on V_1. Modify the method of Exercise 3.5.5 to show

$$\lambda_{3j}^D \leq 5\lambda_j^D \leq \lambda_{3j+3}^D$$

and hence

$$3\rho^D(x) \geq \rho^D(5x) \geq 3\rho^D(x) - 3.$$

This gives another proof of (3.5.5) that does not use spectral decimation.

3.6 INTEGRALS INVOLVING EIGENFUNCTIONS

To complete the story of spectral decimation we need to be able to compute inner products of eigenfunctions. As we mentioned in Section 3.1, the integrals $\int_0^1 \cos^2 \pi kx\,dx$ and $\int_0^1 \sin^2 \pi kx\,dx$ may be determined by spectral decimation, as well as by the standard calculus computations. In this section we will do the analogous computation on SG. Of course the usual elementary arguments show that Dirichlet eigenfunctions with different eigenvalues are orthogonal. The same is true for two Neumann eigenfunctions. So, the real issue is how to compute inner products within eigenspaces. Since the eigenspaces may have high multiplicity, this is an important issue. We can't just leave it to numerical integration.

Actually, the method we describe works for either Dirichlet or Neumann boundary conditions, but we assume first that we have Dirichlet boundary conditions. The idea is to use the discrete approximations to the integral given by (1.2.17) and to see how they change with m. Let λ be a fixed eigenvalue with generation of birth m_0 and discrete eigenvalues $\{\lambda_m\}_{m\geq m_0}$. If u and v are any two λ-eigenfunctions, we consider

$$(3.6.1) \qquad \langle u, v\rangle_m = 3^{-m}\left(\frac{2}{3}\sum_{x\in V_m\setminus V_0} u(x)v(x) + \frac{1}{3}\sum_{x\in V_0} u(x)v(y)\right)$$

$$= 3^{-m-1}\sum_{|w|=m}\sum_i u(F_w q_i)v(F_w q_i)$$

for $m \geq m_0$ as discrete inner products that converge to the true inner product

$$(3.6.2) \qquad \langle u, v\rangle = \int_K u(x)v(x)d\mu(x).$$

We want to understand how these discrete inner products change as we vary m. Specifically, we will show that $\langle u, v\rangle_m$ is just a multiple of $\langle u, v\rangle_{m-1}$, and the multiple is determined by λ_m. For simplicity we take $u = v$ (the general case follows by polarization). There are two ideas in the argument. The first is to use (3.2.7) on each $(m-1)$-cell. So, if $|w| = m - 1$, the contribution to $\langle u, u\rangle_{m-1}$ from the cell $F_w K$ is just

$$(3.6.3) \qquad 3^{-m} \sum_i |u(F_w q_i)|^2,$$

while the contribution to $\langle u, u\rangle_m$ from the three m-cells in $F_w K$ is

$$(3.6.4) \qquad 3^{-m-1}\left(\sum_i |u(F_w q_i)|^2 + 2\sum_i |u(F_w p_i)|^2\right),$$

where $\{p_i\}$ are the vertices in $V_1 \setminus V_0$. Then (3.2.7) allows us to express $u(F_w p_i)$ as a linear combination of $\{u(F_w q_i)\}$, namely

$$(3.6.5) \qquad u(F_w p_i) = a_m (u(F_w q_{i+1}) + u(F_w q_{i-1})) + b_m u(F_w q_i)$$

for

$$(3.6.6) \qquad a_m = \frac{4 - \lambda_m}{(2 - \lambda_m)(5 - \lambda_m)}, \qquad b_m = \frac{2}{(2 - \lambda_m)(5 - \lambda_m)}.$$

When we substitute (3.6.5) in (3.6.4) we obtain

$$(3.6.7) \qquad 3^{-m-1}\left((1 + 2(2a_m^2 + b_m^2)) \sum_i |u(F_w q_i)|^2 \right.$$
$$\left. + (2a_m^2 + 4a_m b_m) \sum_{i \neq j} u(F_w q_i) u(F_w q_j)\right).$$

This introduces cross-terms that are not present in (3.6.3), so we need another idea to get rid of them; namely, we sum over all $(m-1)$-cells and then use the Δ_{m-1} eigenvalue equation. The sum of (3.6.7) over all $(m-1)$-cells yields

$$(3.6.8) \qquad \langle u, u\rangle_m = \frac{1}{3}\left(1 + 4a_m^2 + 2b_m^2\right) \langle u, u\rangle_{m-1}$$
$$+ 3^{-m-1}\left(2a_m^2 + 4a_m b_m\right) \sum_{x \underset{m-1}{\sim} y} u(x) u(y)$$

since every edge in Γ_{m-1} occurs exactly once in (3.6.7). On the other hand, the discrete Gauss–Green formula ((2.2.7) with $u = v$) tells us that

$$(3.6.9) \qquad E_{m-1}(u) = \sum_{x \underset{m-1}{\sim} y} |u(x) - u(y)|^2$$
$$= -\sum_{V_{m-1}\setminus V_0} u(x)\Delta_{m-1}u(x) = \lambda_{m-1} \sum_{x \in V_{m-1}\setminus V_0} |u(x)|^2.$$

Since

(3.6.10) $$\sum_{\substack{x \underset{m-1}{\sim} y}} |u(x) - u(y)|^2 = 4 \sum_{V_{m-1} \backslash V_0} |u(x)|^2 - 2 \sum_{\substack{x \underset{m-1}{\sim} y}} u(x)u(y),$$

we obtain

(3.6.11) $$\sum_{\substack{x \underset{m-1}{\to} \sim y}} u(x)u(y) = \frac{1}{2}(4 - \lambda_{m-1}) \sum_{V_{m-1} \backslash V_0} |u(x)|^2.$$

We substitute this into (3.6.8) to obtain

(3.6.12) $\langle u, u \rangle_m = \dfrac{1}{3} \left(1 + 4a_m^2 + 2b_m^2 + (4 - \lambda_{m-1})\left(a_m^2 + 2a_m b_m\right)\right) \langle u, u \rangle_{m-1}.$

This is an identity of the desired form, but the coefficient seems quite incomprehensible. Using (3.6.6) and $\lambda_{m-1} = \lambda_m(5 - \lambda_m)$, we may express the coefficient entirely in terms of λ_m. After some elementary but complicated algebraic manipulations, we obtain

(3.6.13) $$\langle u, u \rangle_m = b(\lambda_m) \langle u, u \rangle_{m-1}$$

for

(3.6.14) $$b(t) = \frac{\left(1 - \frac{1}{6}t\right)\left(1 - \frac{2}{5}t\right)}{\left(1 - \frac{1}{5}t\right)\left(1 - \frac{1}{2}t\right)}.$$

We leave the details to the exercises. The same result holds if u satisfies Neumann boundary conditions, but again we leave the details to the exercises.

We note that

(3.6.15) $$b(t) = 1 + \frac{2}{15}t + O(t^2) \quad \text{as } t \to 0.$$

Since $\lambda_m = O(5^m)$ we may iterate (3.6.13) and then take the limit to obtain

(3.6.16) $$\langle u, u \rangle = \left(\prod_{m=m_0+1}^{\infty} b(\lambda_m) \right) \langle u, u \rangle_{m_0}.$$

We can simplify the infinite product in (3.6.16) by introducing the special function

(3.6.17) $$\tilde{b}(t) = \prod_{k=1}^{\infty} b((\varphi_-)^k t).$$

Since $\lambda_{m_1+k} = (\varphi_-)^k(\lambda_{m_1})$, where m_1 is the generation of fixation, we can combine all the factors corresponding to $m > m_1$ to obtain

(3.6.18) $$\langle u, u \rangle = \left(\prod_{m=m_0+1}^{m_1} b(\lambda_m) \right) \tilde{b}(\lambda_{m_1}) \langle u, u \rangle_{m_0}.$$

We could also start at level m_1 and obtain

(3.6.19) $$\langle u, u \rangle = \tilde{b}(\lambda_{m_1}) \langle u, u \rangle_{m_1}.$$

This means that if we have an orthonormal basis for the eigenfunctions of Δ_{m_1}, then we may obtain an orthonormal basis for the bottom of the spectrum of Δ by multiplying each eigenfunction by $\tilde{b}(\lambda_{m_1})^{-1/2}$ (with the exception of the top eigenvalue $\lambda_{m_1} = 6$, where we take $b(3)\tilde{b}(3)$ in place of $\tilde{b}(6)$).

EXERCISES

3.6.1. Find a formula for $\int_K u\,d\mu$ for any eigenfunction u (no boundary conditions needed) in terms of the discrete approximation (1.2.17).

3.6.2. Show that (3.6.12) reduces to (3.6.13). (Hint: $4 - \lambda_{m-1} = (1 - \lambda_m)(4 - \lambda_m)$.)

3.6.3.* Show that (3.6.12) also holds for Neumann eigenfunctions. (Hint: You will need to use (2.3.5) in place of (2.2.7)), and (3.3.1).)

3.6.4. Show that $\int_0^1 \sinh x \sinh(1 - x)dx$ cannot be a multiple of the discrete inner product $\langle \sinh x, \sinh(1 - x) \rangle_0$. This shows why we can't expect (3.6.12) to hold without some boundary conditions.

3.6.5. Prove that the infinite product in (3.6.16) converges. Moreover, show that the function $\tilde{b}(t)$ is continuous and increasing on $[0, 5]$.

3.7 NOTES AND REFERENCES

The material in Section 3.1 should be well-known, but it may not have been written down before. The method of spectral decimation as presented in Sections 3.2–3.4 is from [Shima 1991] and [Fukushima & Shima 1992]. They cite earlier work in the physics literature, [Rammal & Toulouse 1983] and [Rammal 1984], as inspiration. Exercise 3.3.10 is from [Gibbons et al. 2001]. Graphs of eigenfunctions may be found in [Dalrymple et al. 1999], which gives an explicit basis for the 5-series and 6-series eigenspaces. The basis for the 5-series described in Section 3.3 is due to Kigami (personal communication). The prevalence of D–N eigenvalues discussed in Section 3.4 is from [Kigami 1998]. The spectral asymptotics in Section 3.5 is due to [Kigami & Lapidus 1993]. Their proof is the one in Exercise 3.5.7. The proof in the text is new. The integral formulas in Section 3.6 are from [Oberlin et al. 2003].

Chapter Four

Postcritically Finite Fractals

4.1 DEFINITIONS

In this chapter we would like to describe a class of self-similar fractals for which it is possible to extend the construction of a self-similar energy given in Chapter 1 for SG. In order for this approach to have a chance of success, the fractal must subdivide into cells of smaller and smaller size, and the cells must intersect at isolated points. In this section we will describe a class of self-similar fractals that have this structure. In the next section we will see what needs to be done to construct the energy. In fact, one does not have to go to any new fractals to come up against some deep problems. There are many different self-similar energies on SG (not having the full dihedral-3 symmetry), for example. This is even true for I, but in that case they are all equivalent after a change of variable.

We will not attempt to describe the theory in greatest generality, but instead make some additional assumptions that greatly simplify the exposition. We begin by assuming that the fractal K is the invariant set for a finite iterated function system (i.f.s.) of contractive similarities in some Euclidean space $\mathbb{R}^n$. We denote these mappings $\{F_i\}_{i=1,\ldots,N}$. Then K is the unique nonempty compact set satisfying

$$(4.1.1) \qquad\qquad K = \bigcup_i F_i K.$$

We are really interested only in the topological properties of K, so the exact contraction ratios of the mappings F_i will not play a role in our discussion. We will only consider connected sets K, but we want the connectedness to be very unstable. In particular, we want the cells $F_i K$ in (4.1.1) to intersect each other in only finitely many points. Thus, by removing these finite number of points, we make the fractal disconnected. Such fractals are often referred to as "finitely ramified", but we will not make this a formal definition. Instead, we define a somewhat more structured class of fractals.

DEFINITION 4.1.1 K is called *postcritically finite* (pcf) if K is connected, and there exists a finite set $V_0 \subseteq K$ called the *boundary*, such that

$$(4.1.2) \qquad F_w K \cap F_{w'} K \subseteq F_w V_0 \cap F_{w'} V_0 \quad \text{for } w \neq w' \text{ with } |w| = |w'|,$$

with the intersection disjoint from V_0. Moreover, we require that each boundary point is the fixed point of one of the mappings $\{F_i\}$. Without loss of generality $V_0 = \{q_1, \ldots, q_{N_0}\}$ for $N_0 \leq N$ and

$$(4.1.3) \qquad\qquad F_i q_i = q_i \quad \text{for } i \leq N_0.$$

In the case of SG we had $N_0 = N$, but we are not requiring that here. Also, the choice of boundary is not necessarily unique, as we will see in some examples. The condition (4.1.2) imposes a structure on the nature of the intersections of cells that persists at all levels. The condition (4.1.3) is made to simplify the exposition.

It is straightforward to construct a sequence of graphs $\Gamma_0, \Gamma_1, \ldots$ as we did for SG. We take Γ_0 to be the complete graph on V_0. Having constructed Γ_{m-1} with vertices V_{m-1}, we set

$$(4.1.4) \qquad V_m = \bigcup_i F_i V_{m-1}$$

and define the edge relation $x \underset{m}{\sim} y$ to hold if and only if we can write $x = F_i x'$ and $y = F_i y'$ with $x' \underset{m-1}{\sim} y'$. Equivalently, $x \underset{m}{\sim} y$ if there exists a word w of length $|w| = m$ (words are now in the letters $\{1, \ldots, N\}$) such that $x, y \in F_w V_0$. The condition (4.1.3) guarantees that $V_0 \subseteq V_1 \subseteq V_2 \subseteq \cdots$. We write

$$(4.1.5) \qquad V_* = \bigcup_m V_m.$$

We can speak of m-cells, the images $F_w K$ for $|w| = m$, and their boundaries $F_w V_0$. Note that V_m consists of all the boundary points of all the m-cells. Two m-cells can intersect only at boundary points. Points in V_m that belong to two or more m-cells are called *junction points*. The situation is more complex than in the case of SG, where all points in $V_m \setminus V_0$ are junction points, and exactly two cells meet at each junction point. In general, there can be points in $V_m \setminus V_0$ that are not junction points, and more than two m-cells can meet at a junction point, as some examples will show. What is true is that the nature of a vertex does not vary with m. If $x \in V_m$, then $x \in V_{m'}$ for all $m' > m$, and x is a junction point in V_m if and only if it is a junction point in $V_{m'}$, in which case the same number of cells meet at x.

It is clear that

$$(4.1.6) \qquad V_* = \bigcup_m V_m$$

is dense in K, so for continuous functions it suffices to understand their values on V_*. Also, the structure of the graphs Γ_m captures the connectivity properties of the fractal at a given resolution. What is perhaps not so clear is why we insist on taking the initial graph Γ_0 to be the complete graph on V_0. Why not just require that Γ_0 be some connected graph on V_0? It would seem that in some cases, when V_0 is large or possesses some special structure, other choices would be more natural. The reason is rather technical and will not be apparent until the next section. But essentially, you have to make this decision if you want to construct self-similar energy forms.

We now look at some examples.

Example 4.1.1 (The level-3 Sierpinski gasket), SG_3 (Figure 4.1.1). Instead of bisecting the sides of a triangle and retaining three of the four smaller triangles, we trisect the sides and retain six of the nine smaller triangles. So $N = 6$, $N_0 = 3$, V_0 consists of the vertices of the triangle, and V_1 is shown in Figure 4.1.2. Note that all seven vertices in $V_1 \setminus V_0$ are junction points, but the one in the middle intersects three 1-cells. In a similar manner we could define SG_n for any value of $n \geq 2$.

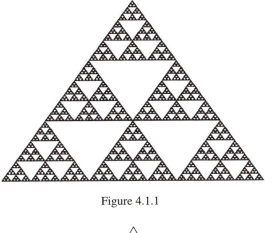

Figure 4.1.1

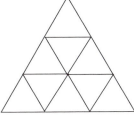

Figure 4.1.2

Figure 4.1.3

Example 4.1.2 (The hexagasket) (Figure 4.1.3). Start with a regular hexagon, and take the six homotheties with contraction ratio $\frac{1}{3}$ and fixed points equal to the vertices of the hexagon. Here $N = 6$ and $N_0 = 6$, V_0 consists of the vertices of the hexagon, and V_1 is shown in Figure 4.1.4. Note that $\#V_1 = 30$, but there are only six junction points, so there are 18 vertices in $V_1 \backslash V_0$ that are not junction points. The boundary of every 1-cell contains exactly two junction points and one

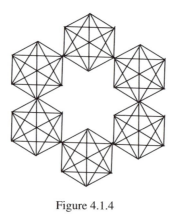

Figure 4.1.4

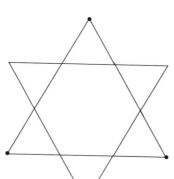

Figure 4.1.5

boundary point. But the situation becomes more diverse when we look at cells of smaller size. Some have two junction points while others have three. This example has dihedral-6 (D_6) symmetry.

Example 4.1.2′ (The hexagasket with smaller boundary). We consider the same fractal, but generated by a slightly different i.f.s. For each F_j in Example 2, we compose it with a rotation through angle $j\pi/3$. Then we may take V_0 to consist of just three of the six vertices, every other one. So now $N_0 = 3$. V_1 takes the shape of a star of David, as shown in Figure 4.1.5. In fact, it is easy to show that the fractal contains all the lines in this star. Since we have broken the symmetry, this example has only D_3 symmetry.

Example 4.1.3 (Lindstrøm snowflake). Take the original six homotheties defining the hexagasket and add one more with fixed point in the center (same contraction ratio). Take V_0 as before, so $N_0 = 6$ and $N = 7$. This example has D_6 symmetry.

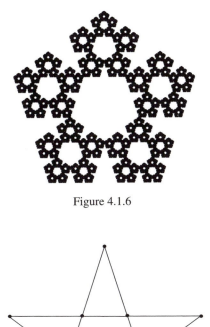

Figure 4.1.6

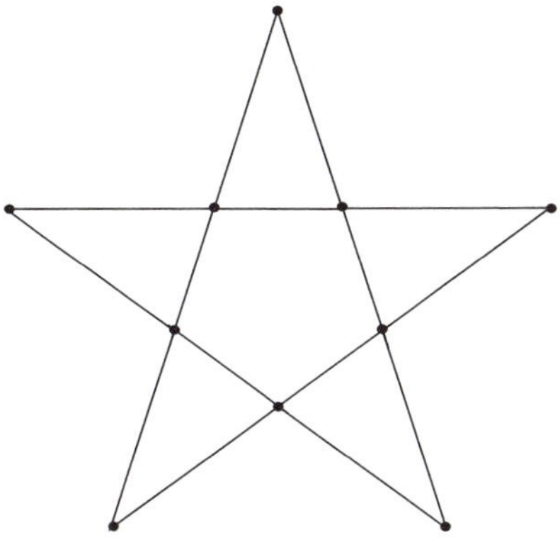

Figure 4.1.7

Example 4.1.4 (Polygaskets). For each n not divisible by 4, start with a regular n-gon and take n homotheties whose fixed points are the vertices of the n-gon, and contraction ratio chosen so that the images of the n-gon just intersect at single points (see the exercises for the exact value). Note that this construction fails when n is divisible by 4, as the best one can achieve is to have the n-gons intersect along edges. Take V_0 to be the set of vertices of the original n-gon. This gives the n-gasket with $N = N_0 = n$ and D_n symmetry. Alternatively, introduce rotations though angle $\frac{2\pi j}{n}$ in each F_j, to obtain the same n-gon. Now we can take for V_0 just three of the n vertices. Figure 4.1.6 shows the pentagasket, and Figure 4.1.7 shows V_1 for the pentagasket. See the exercises for the general case.

Example 4.1.5 (Vicsek set). Consider a square with corners $\{q_1, q_2, q_3, q_4\}$ and center q_5. Choose F_i to be a homothety with contraction ratio $\frac{1}{3}$ and fixed point q_i, $i = 1, \ldots, 5$. Then $N = 5$, $N_0 = 4$ with $V_0 = \{q_1, q_2, q_3, q_4\}$. Figure 4.1.8 shows V_1. This fractal has D_4 symmetry. It is topologically trivial.

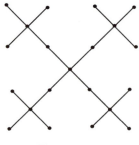

Figure 4.1.8

EXERCISES

4.1.1. Find a formula for the number of vertices in V_m for Example 1.

4.1.2. Sketch V_2 for Example 2′.

4.1.3. Show that all the straight lines in Figure 4.1.5 actually belong to the hexagasket. (Hint: If A denotes this set of lines, show that $A \subseteq \cup F_i A$.)

4.1.4. What does the polygasket construction yield when $n = 4$?

4.1.5. Find the contraction ratio for the n-gasket.

4.1.6. Find a set V_0 of three vertices of the n-gon that will serve as a boundary of the n-gasket.

4.1.7. Prove that the Vicsek set is contractible.

4.2 ENERGY RESTRICTION AND RENORMALIZATION

In this section we assume K is a pcf fractal according to Definition 4.1.1, and we ask if it is possible to construct a self-similar energy using the method described in Sections 1.3 and 1.4 for SG. So the analog of Theorem 1.4.5 would replace the identity (1.4.8) with

$$(4.2.1) \qquad \mathcal{E}(u) = \sum_{i=1}^{N} r_i^{-1} \mathcal{E}(u \circ F_i)$$

for certain resistance renormalization factors r_i satisfying $0 < r_i < 1$. Note that although we have answered this question in the case of SG when all $r_i = \frac{3}{5}$, we could raise it anew for SG with other choices of $\{r_i\}$.

As before, we want $\mathcal{E}$ to be the limit of graph energies $\mathcal{E}_m$ defined on Γ_m. In general, a graph energy will have the form

$$(4.2.2) \qquad \sum_{x \sim y} c(x, y)(u(x) - u(y))^2$$

for some positive conductances $\{c(x, y)\}$ defined on the edges of the graph. We insist on positivity, not just nonnegativity, in order to avoid ending up with degenerate energy forms on the fractal. So we need a sequence $\{c_m(x, y)\}$ of sets of conductances on the edges of the graphs Γ_m, and then we set

$$(4.2.3) \qquad \mathcal{E}_m(u) = \sum_{x \underset{m}{\sim} y} c_m(x, y)(u(x) - u(y))^2.$$

How should we choose these conductances? There are two constraints we would like to satisfy. The first is that the restriction of $\mathcal{E}_m$ to Γ_{m-1} equals $\mathcal{E}_{m-1}$. That means that if u is defined on V_{m-1}, and we consider among all extensions u' of u to V_m the one $\tilde{u}$ that minimizes $\mathcal{E}_m$:

$$(4.2.4) \qquad \mathcal{E}_m(\tilde{u}) \leq \mathcal{E}_m(u') \quad \text{if } \tilde{u} = u' = u \quad \text{on } V_{m-1},$$

then

$$(4.2.5) \qquad \mathcal{E}_m(\tilde{u}) = \mathcal{E}_{m-1}(u).$$

This implies immediately that

$$(4.2.6) \qquad \mathcal{E}_m(u') \geq \mathcal{E}_{m-1}(u)$$

for any extension, so the sequence $\{\mathcal{E}_m(u)\}$ is monotone increasing for any function defined on V_*, and so

$$(4.2.7) \qquad \mathcal{E}(u) = \lim_{m \to \infty} \mathcal{E}_m(u)$$

is well-defined (but may be $+\infty$).

The second constraint is that the sequence of energies must satisfy a version of self-similarity that will produce (4.2.1) in the limit. We may express this simply as

$$(4.2.8) \qquad \mathcal{E}_m(u) = \sum_{i=1}^{N} r_i^{-1} \mathcal{E}_{m-1}(u \circ F_i).$$

Note that if u is defined on V_m, then $u \circ F_i$ is defined on V_{m-1}, so the right side of (4.2.8) makes sense. We may interpret (4.2.8) as an inductive definition. Once we choose $\mathcal{E}_0$, namely

$$(4.2.9) \qquad \mathcal{E}_0(u) = \sum_{i<j} c_{ij}(u(q_i) - u(q_j))^2$$

for positive values of c_{ij}, then (4.2.8) defines $\mathcal{E}_m$ for all m. We may also write $\mathcal{E}_m$ explicitly as

$$(4.2.10) \qquad \mathcal{E}_m(u) = \sum_{|w|=m} \sum_{i<j} r_w^{-1} c_{ij}(u(F_w q_i) - u(F_w q_j))^2,$$

where $r_w = r_{w_1} r_{w_2} \cdots r_{w_m}$.

Since the second constraint effectively determines $\mathcal{E}_m$, we have to go back and ask if the first constraint is satisfied. In other words, if $\mathcal{E}_m$ is defined by (4.2.10) for all m, does (4.2.5) hold? It is straightforward to see that if (4.2.5) holds for $m = 1$, then it holds for all m, since what goes on in each $(m - 1)$-cell is independent of every other cell and is independent of m. So the problem reduces to showing (4.2.5) for $m = 1$. What are the parameters that we must choose? Exactly the renormalization factors $\{r_i\}$ and the initial conductances $\{c_{ij}\}$. Finding the correct values for these parameters is a highly nontrivial task.

It is useful to reformulate the problem slightly. Suppose we are given values for $\{r_i\}$ and $\{c_{ij}\}$ (not necessarily the correct ones). Then we may use them to define

$$(4.2.11) \qquad \mathcal{E}_1(u) = \sum_i \sum_{j<k} r_i^{-1} c_{jk} (u(F_i q_j) - u(F_i q_k))^2$$

on V_1. We may then restrict $\mathcal{E}_1$ to Γ_0, say

$$(4.2.12) \qquad \tilde{\mathcal{E}}_1(u) = \mathcal{E}_1(\tilde{u})$$

for u defined on V_0, where $\tilde{u}$ is the extension of u to V_1 that minimizes $\mathcal{E}_1$. It is easy to see that $\tilde{\mathcal{E}}_1$ must have the form

$$(4.2.13) \qquad \tilde{\mathcal{E}}_1(u) = \sum_{i<j} \tilde{c}_{ij} (u(q_i) - u(q_j))^2$$

for some nonnegative coefficients $\tilde{c}_{ij}$. Note that (4.2.5) for $m = 1$ says that $\tilde{\mathcal{E}}_1 = \mathcal{E}_0$, or $\tilde{c}_{ij} = c_{ij}$ for all $i < j$. However, we would be equally pleased if we could find

$$(4.2.14) \qquad \tilde{c}_{ij} = \lambda c_{ij} \quad \text{for some } \lambda > 0,$$

for then we could go back and correct the renormalization factors to $\{\lambda^{-1} r_i\}$. (Of course we would also want to check that $\lambda^{-1} r_i < 1$ for all i.) This is exactly what happened in the case of SG, where we started with all $c_{ij} = 1$ and all $r_i = 1$, and then found (4.2.14) with $\lambda = \frac{5}{3}$ to conclude that we really want to take all $r_i = \frac{3}{5}$.

We may regard (4.2.14) as a nonlinear eigenvalue problem. Because we are seeking strictly positive solutions, we may regard it as a nonlinear Perron–Frobenius eigenvalue problem. Note that what we are doing is fixing the values of $\{r_i\}$ in $\mathbb{R}_+^N$ (actually the projective version, since multiplying by a positive constant simply changes the eigenvalue) and using this to define the mapping $\{c_{ij}\} \to \{\tilde{c}_{ij}\}$ on $R_+^{N_0(N_0-1)/2}$. (A priori the values of $\tilde{c}_{ij}$ are only known to be nonnegative, but if we get a solution to (4.2.14) with all c_{ij} strictly positive, then we are vindicated.) There is no general theory to which we can turn to provide us with solutions. Since it is an eigenvalue problem, we may always multiply a solution by a positive constant to obtain another solution. To avoid this trivial nonuniqueness, we will normalize the conductances by the condition $\sum_{i<j} c_{ij} = 1$ (or another constant).

We will call $\{c_{ij}\} \to \{\tilde{c}_{ij}\}$ the *renormalization map*, and we will call (4.2.14) the *renormalization equation*. We would like to be able to answer the following questions:

(1) Does there exist a positive solution?
(2) Is there a unique normalized solution?
(3) Is the unique normalized solution attracting? This would mean that we could find the solution in the limit by starting anywhere and iterating the renormalization map, followed by normalizing. (A weaker statement would be that the solution has an open basin of attraction, so that we get it by iteration if we start close enough to it.)

Note that the renormalization map and the answers to the above questions depend on the values of $\{r_i\}$. This leads to two more questions:

(4) Given K, does there exist a choice of $\{r_i\}$ that leads to an affirmative answer to one or more of the above questions?

(5) If so, can we specify exactly which choices of $\{r_i\}$ lead to affirmative answers?

We only have piecemeal answers to these questions. We will discuss some of these answers for some examples in the next section. In particular, it is not known whether or not question (4) has an affirmative answer for every K (with regard to existence). There are known examples where uniqueness fails, but if uniqueness holds, then the solution is attracting. Nevertheless, there are enough examples where explicit solutions are known to make for an interesting theory. In particular, we don't have to worry about what the future might bring. An affirmative answer to question (4) for every K would be a terrific result; but if the answer in general is negative, it will lead to the interesting question of characterizing those fractals for which a solution exists.

For the remainder of this section we will assume that we have a solution of the renormalization problem and derive some simple consequences. We will assume $\lambda = 1$ in (4.2.14), as we may by adjusting the $\{r_i\}$ values. We now want to examine the extension $\tilde{u}$ of u from V_0 to V_1 that minimizes energy. As before, we will call it the *harmonic extension*. First we want to write the linear equations that the values of $\tilde{u}(x)$ must satisfy for $x \in V_0 \backslash V_1$. From (4.2.3) we derive

$$(4.2.15) \qquad \sum_{y \underset{1}{\sim} x} c_1(x, y)(\tilde{u}(x) - \tilde{u}(y)) = 0 \quad \text{for } x \in V_1 \backslash V_0$$

by differentiating with respect to $\tilde{u}(x)$ and setting the derivative equal to zero. Of course we have to read off the values of $c_1(x, y)$ from (4.2.11), namely

$$(4.2.16) \qquad c_1(x, y) = r_i^{-1} c_{jk} \qquad \text{if } x = F_i q_j \quad \text{and} \quad y = F_i q_k$$

(such a representation is equivalent to $x \underset{1}{\sim} y$). It is easy to interpret (4.2.15) as saying that the value $\tilde{u}(x)$ is a weighted average of the values of $\tilde{u}$ at the neighbors y of x. Note that (4.2.15) is a system of inhomogeneous linear equations, with the same number of equations as unknowns ($\#(V_1 \backslash V_0)$), for each specification of u on V_0. Since an energy minimizer clearly exists, the system always has a solution, and by linear algebra it must have a unique solution. So the value of $\tilde{u}(x)$ is a linear combination of the boundary values $u(q_i)$. Since the averaging property implies a maximum principle—$\tilde{u}$ assumes its maximum and minimum value on the boundary—it follows that the coefficients are all nonnegative:

$$(4.2.17) \qquad \tilde{u}(x) = \sum_{i=1}^{N_0} a_{xi} u(q_i) \quad \text{for } a_{xi} \geq 0.$$

This is the analog of the "$\frac{1}{5} - \frac{2}{5}$ rule" for SG. We may collect these coefficients into $N_0 \times N_0$ matrices A_i, $i = 1, \ldots, N$, such that

$$(4.2.18) \qquad (\tilde{u} \circ F_i)|_{V_0} = A_i(u|_{V_0}).$$

These are the *harmonic extension* matrices. In general, these will not be invertible.

It is reasonable to expect that most of the coefficients a_{xi} are strictly positive. If we substitute (4.2.17) into (4.2.11) we will have an explicit formula for $\mathcal{E}_1(\tilde{u})$ that will tell us the coefficients $\tilde{c}_{ij}$ in (4.2.13), in particular that they are all strictly positive. In view of the renormalization equation (4.2.14), this means the original coefficients $\{c_{ij}\}$ are all strictly positive. We built this into our assumptions, of course, but this argument explains why it is necessary to do so. This also explains why we take Γ_0 to be the complete graph on V_0. To take a smaller graph would in essence declare some of the coefficients c_{ij} to be zero, and that would effectively rule out finding a nondegenerate solution to the renormalization equation. We should also point out that there are usually many degenerate solutions, corresponding to graphs on V_0 that do not involve all vertices, or are disconnected. We deliberately want to exclude these. For example, on SG if we take $\mathcal{E}_0(u) = (u(q_1) - u(q_2))^2$ and $r_1 = r_2 = \frac{1}{2}, r_3 = 0$, we would simply obtain the energy of the restriction of u to the interval joining q_1 and q_2.

EXERCISES

4.2.1. Show that for any connected graph and energy given by (4.2.2), u has zero energy if and only if u is constant.

4.2.2. Show that (4.2.5) holds for all $m \geq 1$ if it holds for $m = 1$.

4.2.3. Show that if $\tilde{\mathcal{E}}_1$ is defined by (4.2.12), then it must have the form (4.2.13).

4.2.4.* Write the values of u on V_1 as a column vector $\binom{u_0}{u_1}$, where u_0 lists the values on V_0 and u_1 lists the values on $V_1 \backslash V_0$. Then write $\mathcal{E}_1(u)$ in matrix form $Mu \cdot u$ where

$$M = \begin{pmatrix} T & J^T \\ J & X \end{pmatrix}$$

is symmetric, so that

$$\mathcal{E}_1(u) = T u_0 \cdot u_0 + X u_1 \cdot u_1 + 2 J u_0 \cdot u_1.$$

Show that X is invertible, and for the harmonic extension $\tilde{u}$ we have $\tilde{u}_1 = X^{-1} J u_0$. Thus

$$\mathcal{E}_1(\tilde{u}) = (T - J^T X^{-1} J) u \cdot u.$$

4.2.5. Show that $\mathcal{E}_0$ determines c_{ij} directly by $c_{ij} = \frac{1}{4}(\mathcal{E}_0(u_1) - \mathcal{E}_0(u_2))$, where $u_1(q_i) = -u_1(q_j) = 1$ and $u_1(q_k) = 0$ for $k \neq i$ or j, while $u_2(q_i) = u_2(q_j) = 1$ and $u_2(q_k) = 0$ for $k \neq i$ or j.

4.2.6. Show that any graph energy of the form (4.2.2) has the Markov property $\mathcal{E}(\min\{1, \max\{u, 0\}\}) \leq \mathcal{E}(u)$.

4.2.7.* Show that any nonnegative definite quadratic form on functions on V that annihilates constants and has the Markov property must have the form (4.2.2).

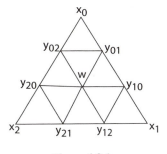

Figure 4.3.1

4.3 EXAMPLES

In this section we explicitly describe some solutions of the renormalization problem and the corresponding harmonic extension algorithm on some of the pcf fractals discussed in Section 4.1. We will use the same numbering as before.

Example 4.3.1 For SG_3, we look for a D_3-symmetric solution, so we choose all r_i equal. Initially we set all $r_i = 1$. We also choose all $c_{jk} = 1$. We label the values of the function $\tilde{u}$ on V_1 as shown in Figure 4.3.1, so x_0, x_1, x_2 are the given values of u on V_0, and $y_{01}, y_{10}, y_{02}, y_{20}, y_{12}, y_{21}, w$ are to be determined. Now the mean value condition (4.2.15) simply says that each of these values is a fair average of the neighboring values, since by symmetry all edges in Γ_1 have the same conductances. So we obtain seven equations:

$$(4.3.1) \qquad 6w - y_{01} - y_{10} - y_{02} - y_{20} - y_{12} - y_{21} = 0,$$

$$(4.3.2) \qquad \begin{cases} 4y_{01} - y_{10} - y_{02} - w = x_0, \\ 4y_{10} - y_{01} - y_{12} - w = x_1, \\ 4y_{02} - y_{01} - y_{20} - w = x_0, \\ 4y_{20} - y_{02} - y_{21} - w = x_2, \\ 4y_{12} - y_{21} - y_{10} - w = x_1, \\ 4y_{21} - y_{12} - y_{20} - w = x_2. \end{cases}$$

Introduce the abbreviations Y and X for the sum of the y and x variables, respectively. Then sum the equations in (4.3.2) to obtain

$$2Y - 6w = 2X,$$

hence $Y = 2X$, $w = \frac{1}{3}X$. With a little more work we find the solution

$$(4.3.3) \qquad \begin{cases} y_{01} = \frac{8}{15}x_0 + \frac{4}{15}x_1 + \frac{3}{15}x_2, \\ y_{10} = \frac{4}{15}x_0 + \frac{8}{15}x_1 + \frac{3}{15}x_2, \\ \text{etc.} \end{cases}$$

Figure 4.3.2 shows the values of a single harmonic function. It is easy to check by inspection that the mean value condition holds at all seven points of $V_1 \backslash V_0$.

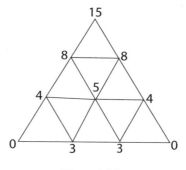

Figure 4.3.2

There is no doubt (by symmetry) that the renormalization equation holds, and by computing the energies for function in Figure 4.3.2 we find that all $r_i = \frac{7}{15}$, because $E_0 = 2(15)^2$ and

$$E_1 = 2 \cdot 7^2 + 2\left(4^2 + 3^2 + 1^2\right) + 2\left(4^2 + 3^2 + 1^2\right) + 2 \cdot 2^2 = 2 \cdot 7 \cdot 15.$$

The matrix A_0 is given by

(4.3.4)
$$\begin{pmatrix} 1 & 0 & 0 \\ \frac{8}{15} & \frac{4}{15} & \frac{3}{15} \\ \frac{8}{15} & \frac{3}{15} & \frac{4}{15} \end{pmatrix}$$

with eigenvalues $1, \frac{7}{15}, \frac{1}{15}$ corresponding to the eigenvectors

(4.3.5)
$$\begin{pmatrix} 1 \\ 1 \\ 1 \end{pmatrix}, \begin{pmatrix} 0 \\ 1 \\ 1 \end{pmatrix}, \begin{pmatrix} 0 \\ 1 \\ -1 \end{pmatrix}.$$

Example 4.3.2′ The hexagasket with $N_0 = 3$ also has D_3 symmetry, so we take all $c_{jk} = 1$, and for simplicity all r_i equal. (We could look at a more general situation with two choices of r_i, one for the cells containing boundary points and one for the other cells, and still maintain D_3 symmetry, but in the end this would not have D_6 symmetry.) We label the values of $\tilde{u}$ on V_1 as in Figure 4.3.3, with x_0, x_1, x_2 denoting the values of V_0. Again, each edge in Γ_1 has the same conductance, so the mean value condition is just that each value is the fair average of its neighbors. In particular, we may eliminate the z-variables since

(4.3.6)
$$\begin{cases} z_{01} = \frac{1}{2} y_{01} + \frac{1}{2} y_{10}, \\ z_{12} = \frac{1}{2} y_{12} + \frac{1}{2} y_{21}, \\ z_{20} = \frac{1}{2} y_{20} + \frac{1}{2} y_{02}. \end{cases}$$

The equation at y_{01} is

(4.3.7)
$$4 y_{01} - y_{10} - y_{02} - z_{01} = x_0,$$

which becomes

(4.3.8)
$$\frac{7}{2} y_{01} - \frac{3}{2} y_{10} - y_{02} = x_0,$$

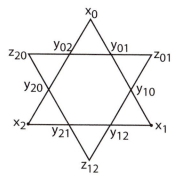

Figure 4.3.3

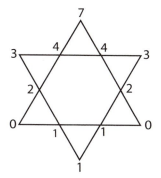

Figure 4.3.4

and similarly for the other y variables. It is easy to check that the solution is

(4.3.9)
$$y_{01} = \frac{4}{7}y_0 + \frac{2}{7}y_1 + \frac{1}{7}y_2, \text{ etc.,}$$

and Figure 4.3.4 shows a typical harmonic function. For this function $E_0 = 2 \cdot 7^2$ and

$$E_1 = 2 \cdot 3^2 + 2\left(2^2 + 1^2 + 1^2\right) + 2\left(2^2 + 1^2 + 1^2\right) + 0 = 2 \cdot 3 \cdot 7,$$

so all $r_i = \frac{3}{7}$. The matrix A_0 is

(4.3.10)
$$\begin{pmatrix} 1 & 0 & 0 \\ \frac{4}{7} & \frac{2}{7} & \frac{1}{7} \\ \frac{4}{7} & \frac{1}{7} & \frac{3}{7} \end{pmatrix}$$

with eigenvalues $1, \frac{3}{7}, \frac{1}{7}$ corresponding to eigenvectors

(4.3.11)
$$\begin{pmatrix} 1 \\ 1 \\ 1 \end{pmatrix}, \begin{pmatrix} 0 \\ 1 \\ 1 \end{pmatrix}, \begin{pmatrix} 0 \\ 1 \\ -1 \end{pmatrix}.$$

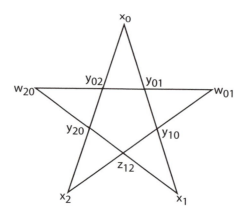

Figure 4.3.5

One very startling observation is that the harmonic function in Figure 4.3.4 has the same values at the boundary points of the bottom cell, and hence it will be constant on the whole cell. This is certainly not the kind of behavior one would expect of a harmonic function! On the other hand, we can give a simple symmetry argument to show why this has to be so: Once the values of y_{12} and y_{21} are equal, then z_{12} must also be equal to this value by (4.3.5). Any harmonic function symmetric with respect to the vertical reflection symmetery must have $y_{12} = y_{21}$. The harmonic extension matrix associated with this cell cannot be invertible, since the constant function has the same values on this cell.

Example 4.3.3 Consider the pentagasket with $N_0 = 3$. In this case we only have a single reflection symmetry, so the initial energy may have the form

(4.3.12) $$E_0(u) = a((x_0 - x_1)^2 + (x_0 - x_2)^2) + b(x_1 - x_2)^2.$$

Without loss of generality we may take $a = 1$ (multiply E_0 by a constant), but we cannot specify b in advance. For simplicity we look for a solution with all r_i equal (initially $r_i = 1$). Then

(4.3.13)
$$\begin{aligned}
E_1(\tilde{u}) = &(x_0 - y_{01})^2 + (x_0 + y_{02})^2 + b(y_{01} - y_{02})^2 \\
&+ (w_{01} - y_{01})^2 + (w_{01} - y_{10})^2 + b(y_{01} - y_{10})^2 \\
&+ (w_{20} - y_{02})^2 + (w_{20} - y_{20})^2 + b(y_{02} - y_{20})^2 \\
&+ (x_1 - y_{10})^2 + (x_1 - z_{12})^2 + b(y_{10} - z_{12})^2 \\
&+ (x_2 - y_{20})^2 + (x_2 - z_{12})^2 + b(z_{20} - z_{12})^2.
\end{aligned}$$

This leads to a linear system of mean value equations for the seven variables $y_{01}, y_{10}, y_{02}, y_{20}, z_{12}, v_{01}, w_{20}$. These equations involve the parameter b, so the solutions depend linearly on b. If we substitute the solutions into (4.3.13) and simplify, we will obtain an expression of the same form as (4.3.12) with new parameters a', b'. The renormalization equation is just $\frac{b'}{a'} = b$. This turns out to be a quadratic equation in b with one positive root. This determines the form of E_0, and the previous solution of the mean value equations (for this choice of b) gives the harmonic extension algorithm. Also, $r_i = \frac{1}{a'}$.

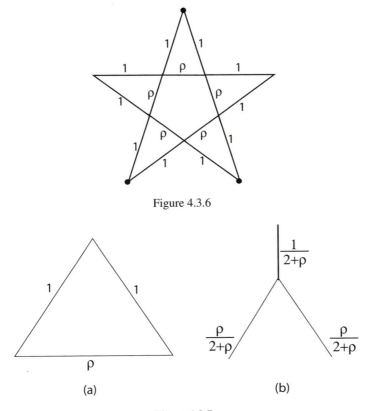

Figure 4.3.6

Figure 4.3.7

The details are messy enough that we try instead the electric network algorithm described in Section 1.5. Since we want to work with resistances rather than conductances, we write $\rho = \frac{1}{b}$. Then E_1 corresponds to the electric network shown in Figure 4.3.6, with resistances of each edge marked. To find the equivalent network with just the three marked vertices, we use the $\Delta - Y$ equivalence in Figure 4.3.7.

Using this five times in Figure 4.3.6 we obtain Figure 4.3.8(a). Then, by pruning and adding resistances in series, we obtain Figure 4.3.8(b). Using another $\Delta - Y$ transform yields Figure 4.3.8(c), and adding resistances in series yields Figure 4.3.8(d). The renormalization equations say exactly that this last network must be a multiple of the network in Figure 4.3.7(b), which yields the equation

(4.3.14) $$\frac{4\rho + 5}{8\rho + 5} = \rho \quad \text{or} \quad 8\rho^2 + \rho - 5 = 0.$$

So we find $\rho = \frac{\sqrt{161}-1}{16}$ as the unique positive solution. Thus $b = \frac{1}{\rho} = \frac{\sqrt{160}+1}{10}$ and all

$$r_i = \frac{5}{8\rho + 5} = \frac{\sqrt{161}-9}{8} \approx .46107.$$

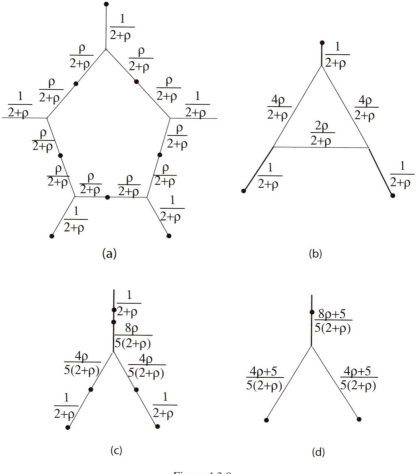

Figure 4.3.8

We conclude this section by analyzing the family of all self-similar energies on SG that have a single bilateral symmetry. So we require $r_1 = r_2$ and

(4.3.15) $\mathcal{E}_0(u) = (u(q_0) - u(q_1))^2 + (u(q_0) - u(q_2))^2 + b(u(q_1) - u(q_2))^2$

for some $b > 0$. Again we write $\rho = \frac{1}{b}$ and we initially take $(r_0, r_1, r_2) = (s, 1, 1)$. The initial network and its $\Delta - Y$ transform is then identical to the previous example, shown in Figure 4.3.7. The level-1 network is shown in Figure 4.3.9. In Figure 4.3.10 we show the transformation of this network, first using the $\Delta - Y$ transform three times to get (a), then adding resistances in series to get (b), then using the $\Delta - Y$ transform to get (c), and then adding resistances in series to get (d). The renormalization equation says that Figure 4.3.10(d) must be a multiple of Figure 4.3.7(b). This leads to the equation

$$2(2s\rho + \rho + 2) = (s\rho + 1)^2 + 2s(s\rho + \rho + 1),$$

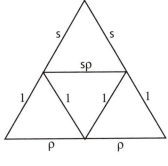

Figure 4.3.9

which is equivalent to

(4.3.16) $s^2\rho^2 + 2s^2\rho + 2s - 2\rho - 3 = 0$

or

(4.3.17) $s^2 + 2s^2b + 2sb^2 - 2b - 3b^2 = 0$

in terms of the conductance b. If (4.3.17) holds, then the corrected values of the r_i are given by

(4.3.18) $(r_0, r_1, r_2) = \dfrac{1+s+b}{1+2s+2b}(s, 1, 1).$

Note that (4.3.17) is a quadratic equation in b for each fixed s, and a quadratic equation in s for each fixed b. These equations have at most one positive solution, so each variable determines the other; explicitly,

(4.3.19) $b = \dfrac{s^2 - 1\sqrt{(s^2-1)^2 + s^2(3-2s)}}{3-2s}$

and

(4.3.20) $s = \dfrac{-b^2 + \sqrt{b(b+1)(b^2+5b+2)}}{1+2b}.$

However, an examination of (4.3.19) reveals that the condition

(4.3.21) $s < \dfrac{3}{2}$

must be imposed in order for the solution to be positive. On the other hand, all positive values of b are allowed. This shows that if we are not careful about the choice of the r factors, the renormalization problem may have no solutions. On the other hand, when a solution exists, it is unique.

EXERCISES

These are best done using a computer algebra package, such as Maple or MATHEMATICA.

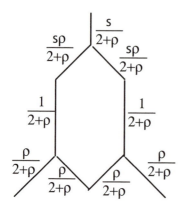

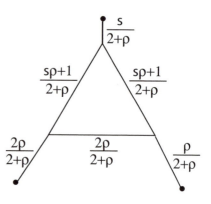

(a)   (b)

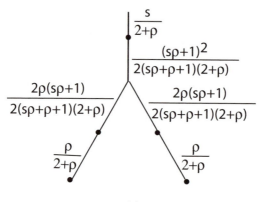

(c)

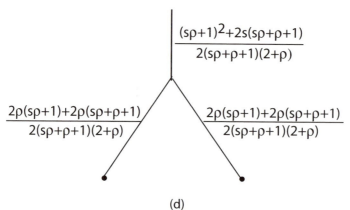

(d)

Figure 4.3.10

4.3.1.* Solve the renormalization problem for the hexagasket with six boundary points as described in Example 2 and full D_6 symmetery, with all $r_i = \frac{3}{7}$. Show that this yields the same energy as before. Find the 6×6 harmonic extension matrices.

4.3.2.* Solve the renormalization problem for the pentagasket with five boundary points and full D_5 symmetry, with all $r_i = \frac{\sqrt{161}-9}{8}$. Show that this yields the same energy as before. Find the 5×5 harmonic extension matrices.

4.3.3.* Find the harmonic extension matrices for the energies on SG with bilateral symmetry.

4.3.4.* For the Vicsek fractal, choose initial conductances equal to 1 along the sides of the square and equal to b and $\frac{1}{b}$ along the two diagonals, for any positive b. Verify that this is a solution to the renormalization problem with all $r_i = \frac{1}{3}$. Find the 4×4 harmonic extension matrices. This shows that solutions to the renormalization problem need not be unique.

4.4 LAPLACIANS

In this section we describe how to carry out the constructions in Chapters 1 and 2 once we have a solution to the renormalization problem. So we assume that K is a pcf fractal, $\mathcal{E}_0$ is given by (4.2.9) for positive c_{ij}, $\mathcal{E}_m$ is given by (4.2.8) or equivalently (4.2.10) for $\{r_i\}$ with $0 < r_i < 1$, and $\tilde{\mathcal{E}}_1 = \mathcal{E}_0$ (so the eigenvalue λ in (4.2.14) is 1); therefore (4.2.5) holds for all m. It follows that $\{\mathcal{E}_m(u)\}$ is always monotone increasing, so

$$(4.4.1) \qquad\qquad \mathcal{E}(u) = \lim_{m \to \infty} \mathcal{E}_m(u)$$

exists, and we define dom $\mathcal{E}$ to be the space of functions for which $\mathcal{E}(u) < \infty$.

Let r denote $\max_i r_i$, so $r < 1$. Then the arguments leading up to (1.4.3) remain valid, and again we may interpret this as a continuity statement, since if x and y are close enough in the Euclidean metric they must belong to the same or adjacent m-cells. (The quantitative statement will depend on the contraction ratios of the i.f.s.) All the results of Section 1.4 are valid, with essentially the same proof: All functions in dom $\mathcal{E}$ are continuous, dom $\mathcal{E}$/constants forms a Hilbert space, piecewise harmonic splines are dense in dom $\mathcal{E}$, and the self-similar identity (4.2.1) holds.

We define the effective resistance metric by (1.6.1) as before. We have already verified in Section 1.6 that it is a metric for any network. Then the Hölder continuity estimate (1.6.2) follows immediately from the definition. We claim that an m-cell $R_w K$ has diameter on the order of r_w in the resistance metric. Suppose x and y are points on the boundary of $F_w K$, say $x = F_w q_i$ and $y = F_w q_j$. Consider the piecewise harmonic spline $\psi_y^{(m)}$ satisfying $\psi_y^{(m)}(z) = \delta_{yz}$ for $z \in V_m$. Then

$$(4.4.2) \qquad\qquad R(x, y) \geq \mathcal{E}\left(\psi_y^{(m)}\right)^{-1} = cr_w$$

for $c = \left(\sum_k c_{kj}\right)^{-1}$.

On the other hand, $u(x) = 0$ and $u(y) = 1$ imply $\mathcal{E}(u) \geq \mathcal{E}_m(u) \geq c_{ij}(r_w)^{-1}$, so we obtain the upper bound

$$(4.4.3) \qquad\qquad R(x, y) \leq c_{ij}^{-1} r_w.$$

Then by a chaining argument we obtain a similar upper bound for any x and y in $F_w K$ (with a different constant). This is certainly enough to conclude that the resistance metric determines the same topology as the Euclidean metric. In general it will not be true that it is equivalent to a power of the Euclidean metric.

We may define a Laplacian Δ_μ for any regular measure μ by Definition 2.1.1. Theorem 2.1.2 holds using the same proof. If μ is a self-similar measure with weights $\{\mu_i\}$, then Δ_μ satisfies the scaling identity (2.1.8). To find a pointwise formula for the Laplacian we need to change notation slightly and consider a graph Laplacian Δ_m on Γ_m based on the conductances in (4.2.3). Thus

$$(4.4.4) \qquad\qquad \Delta_m u(x) = \sum_{y \underset{m}{\sim} x} c_m(x, y)(u(y) - u(x)).$$

Note that the r factors are already incorporated into (4.4.4), so the analog of (2.2.4) will be

$$(4.4.5) \qquad\qquad \Delta_\mu u(x) = \lim_{m \to \infty} \left(\int_K \psi_x^{(m)} d\mu \right)^{-1} \Delta_m u(x).$$

The analog of Theorem 2.2.1 holds with essentially the same proof, so the uniform limit existing in (4.4.5) is equivalent to the weak formulation of the Laplacian. Another way of looking at it is that a Laplacian on a graph is determined by assigning weights to both the edges and the vertices. In Section 2.2 we simply took all these weights to be 1. In (4.4.4) we used the conductances to weight the edges, but we still took the weights on the vertices to be all 1. On the right side of (4.4.5) we assign the weight $\int_K \psi_x^{(m)} d\mu$ to the vertex x, and it is this graph Laplacian that approximates Δ_μ without any renormalization factors.

The definition of the normal derivative at a boundary point q_i is

$$(4.4.6) \qquad\qquad \partial_n u(q_i) = \lim_{m \to \infty} r_i^{-m} \sum_{j \neq i} c_{ij}(u(q_i) - u(F_i^m q_j)).$$

Note that the points $\{F_i^m q_j\}$ are the neighbors of q_i in V_m, and $\{r_i^{-m} c_{ij}\}$ are the conductances of the edges joining q_i to its neighbors. With this definition, the analog of Theorem 2.3.2 is valid: normal derivatives exist for all $u \in \operatorname{dom} \Delta_\mu$ and the Gauss–Green formula (2.3.4) holds. The proof is essentially the same. For harmonic functions you do not have to take the limit in (4.4.6), so

$$(4.4.7) \qquad\qquad \partial_n h(q_i) = \sum_{j \neq i} c_{ij}(u(q_i) - u(q_j)).$$

To localize the definition of the normal derivative just use

$$(4.4.8) \qquad\qquad \partial_n u(F_w q_i) = r_w^{-1} \partial_n (u \circ F_w)(q_i).$$

Then the localized versions of the Gauss–Green formula (2.3.13) and (2.3.14) continue to hold. The matching condition for normal derivatives at any point x in

$V_m \backslash V_0$ states that for $u \in \mathrm{dom}\,\Delta_\mu$, the sum of all the normal derivatives at x (with respect to each m-cell that has x as a boundary point) must be zero. If x is a junction point the sum will have more than one term (but sometimes more than two). However, in some cases there are nonjunction points in $V_m \backslash V_0$. At these points the matching condition says that the normal derivative must be zero! In all cases the observation is that the sum of the normal derivatives is just

$$(4.4.9) \qquad\qquad\qquad \lim_{m \to \infty} (-\Delta_m u(x)),$$

so the existence of the limit in (4.4.5) implies the vanishing of (4.4.9).

The symmetric version of the Gauss–Green formula, Theorem 2.4.1, continues to hold, and Theorem 2.4.2 holds (just delete the r^{-m} factor on the left side of (2.4.5), since it is already included in $\Delta_m u(x)$). The gluing theorem (Theorem 2.5.1) holds with the above form of the matching conditions.

In some of the examples discussed in Section 4.3, we had a choice of a smaller or larger boundary, specifically the hexagasket and pentagasket. We constructed the energy based on a boundary of just three points, but the more natural boundary has six (respectively, five) points. In the exercises we indicated that the same energy exists with the larger boundary. But the dimension of the space of harmonic functions is equal to the number of boundary points. Of course the definition of harmonic function depends on the choice of boundary, since these are the points where we fix the value before minimizing energy. But now we can give a more precise description of what is going on, since the extra points in the larger boundary are nonjunction points in $V_1 \backslash V_0$ with respect to the smaller boundary. So the harmonic functions with respect to the smaller boundary satisfy Neumann boundary conditions at these extra points. The Neumann conditions exactly explain the drop in the dimension of the space of harmonic functions.

Next we discuss the construction of the Green's function. As in Section 2.6 we are seeking a function $G(x, y)$ such that

$$(4.4.10) \qquad\qquad u(x) = \int_K G(x, y) f(y) d\mu(y)$$

solves $-\Delta u = f$ with Dirichlet boundary conditions $v\big|_{V_0} = 0$. We will have

$$(4.4.11) \qquad\qquad G(x, y) = \lim_{M \to \infty} G_M(x, y)$$

for

$$(4.4.12) \qquad\qquad G_M(x, y) = \sum_{m=0}^{M} \sum_{|w|=m} r_w \Psi\left(F_w^{-1}x, F_w^{-1}y\right)$$

with

$$(4.4.13) \qquad\qquad \Psi(x, y) = \sum_{z,z' \in V_1 \backslash V_0} g(z, z') \psi_z^{(1)}(x) \psi_{z'}^{(1)}(y),$$

with $g(z, z')$ to be determined. The goal is to show

$$(4.4.14) \qquad \mathcal{E}(G_M(\cdot, y), v) = \sum_{z \in V_{M+1} \backslash V_0} v(z) \psi_z^{(M+1)}(y) \quad \text{for } v \in \mathrm{dom}_0 \mathcal{E},$$

which is identical to (2.6.17), for then the same proof as in Section 2.6 applies.

First we examine (4.4.14) when $M = 0$. Then $G_0(x, y) = \Psi(x, y)$. We observe that

$$(4.4.15) \qquad \mathcal{E}\left(\psi_z^{(1)}, v\right) = \mathcal{E}_1\left(\psi_z^{(1)}, v\right) = -\Delta_1 v(z).$$

Thus we have

$$(4.4.16) \qquad \mathcal{E}(G_0(\cdot, y), v) = \sum_{z, z' \in V_1 \backslash V_0} g(z, z')\psi_z^{(1)}(y)(-\Delta_1 v(z')),$$

and hence (4.4.14) is equivalent to

$$(4.4.17) \qquad \sum_{V_1 \backslash V_0} g(z, z')(-\Delta_1 v(z')) = v(z) \quad \text{for } z \in V_1 \backslash V_0.$$

In other words, $g(z, z')$ is just the inverse matrix to the operator $-\Delta_1$ on $V_1 \backslash V_0$ (with all functions vanishing on V_0). We know that $-\Delta_1$ has trivial kernel, so by linear algebra it is invertible, and since $-\Delta_1$ is a positive operator we know that $g(z, z')$ are all nonnegative (usually strictly positive). In Section 2.6 we computed $g(z, z')$ for SG explicitly, and for any particular example it is possible to compute it explicitly, although the computations may be complicated.

Then we prove (4.4.14) in general by induction. For the sake of clarity we just do the case $M = 1$. Then

$$(4.4.18) \qquad G_1(x, y) = G_0(x, y) + \sum_j \sum_{z, z' \in V_1 \backslash V_0} r_j g(z, z')\psi_{F_j z}^{(2)}\psi_{F_j z'}^{(2)}(y).$$

Also we have

$$(4.4.19) \qquad \mathcal{E}\left(\psi_{F_j z}^{(2)}, v\right) = \mathcal{E}_2\left(\psi_{F_j z}^{(2)}, v\right) = -\Delta_2 v(F_j z).$$

Using (4.4.18), (4.4.19), and the $M = 0$ case of (4.4.14) we obtain

$$(4.4.20) \qquad \begin{aligned} \mathcal{E}(G_1(\cdot, y), v) &= \sum_{z \in V_1 \backslash V_0} v(z)\psi_z^{(1)}(y) \\ &+ \sum_j \sum_{z, z' \in V_1 \backslash V_0} r_j g(z, z')\psi_{F_j z}^{(2)}(y)(-\Delta_2 v(F_j z')). \end{aligned}$$

It is clear that the right side of (4.4.20) is a linear combination of $v(z)$ for $z \in V_2 \backslash V_0$, so we need to identify the coefficient of $v(z)$ with $\psi_z^{(2)}(y)$. We give a separate argument for $z \in V_2 \backslash V_1$ and $z \in V_1 \backslash V_0$. In the first case $z \in F_j(V_1 \backslash V_0)$ for a unique choice of j. Then the argument is essentially the same as in the $M = 0$ case, with the factor r_j cancelling the r_j^{-1} factor in the definition of Δ_2. In the second case the coefficient of $v(z)$ is

$$(4.4.21) \qquad \psi_z^{(1)}(y) - \sum_{z', z'' \in V_1 \backslash V_0} g(z'', z')c(F_j z', z)\psi_{F_j z''}^{(2)}(y).$$

It is clear that (4.4.21) is a piecewise harmonic spline of level 2, so it suffices to show that it agrees with $\psi_z^{(2)}(y)$ for $y \in V_2$. For $y = z$, the first term in (4.4.21) is 1 and the other terms are all 0, so it agrees with $\psi_z^{(2)}(z)$. The only other places where it is nonzero are the points of the form $F_j z''$ for $z'' \in V_1 \backslash V_0$ that are neighbors of z. If $y = F_j z''$, then the value of (4.4.21) is

$$(4.4.22) \qquad \psi_z^{(1)}(F_j z'') - \sum_{z' \in V_1 \backslash V_0} g(z'', z') r_j c(F_j z', z).$$

Note that $r_j c(F_j z', z) = c(z', \tilde{z})$ if $z = F_j \tilde{z}$. The value of $\psi_z^{(1)}(F_j z'')$ is given by the harmonic extension algorithm. Using the relationship between the harmonic extension algorithm and the $g(z, z')$ matrix, it can be shown that (4.4.22) vanishes. We leave the details to the exercises. Thus the analog of Theorem 2.6.1 holds.

Next we consider the local behavior of a harmonic function in a neighborhood of a boundary point q_i. By the maximum principle it suffices to understand the values on $F_i^m V_0$, the boundary of the cells $F_i^m K$. Instead of working with the full N_0-dimensional space $\mathcal{H}_0$ of harmonic functions, we consider the $(N_0 - 1)$-dimensional space $\mathcal{H}_0'$ of functions vanishing at q_i, a complementary space to the constants. Such functions are determined by their values on $V_0 \backslash \{q_i\}$, and we let A_i' denote the $(N_0 - 1) \times (N_0 - 1)$ matrix representing the transformation

$$h \big|_{V_0 \backslash \{q_i\}} \to h \circ F_i \big|_{V_0 \backslash \{q_i\}}.$$

By the maximum principle this matrix has positive entries, so by the Perron-Frobenius theory it has a positive eigenvalue of multiplicity 1 with positive entry eigenvector, which realizes the spectral radius. We claim that this eigenvalue is exactly r_i. Indeed, if h is the harmonic function corresponding to any eigenvector with eigenvalue λ, then $h(q_i) = 0$ and

$$(4.4.23) \qquad \partial_n h(q_i) = \lim_{m \to \infty} r_i^{-m} \lambda^m \sum_{j \neq i} (-c_{ij}) h(q_j).$$

So if $\sum c_{ij} h(q_j) \neq 0$, which will be the case if h is positive, then we must have $\lambda \leq r_i$. On the other hand, if $\lambda < r_i$ then $\partial_n h(q_i) = 0$. In the exercises we will see that it is impossible to have $\partial_n h(q_i) = 0$ for all harmonic functions. It follows that the Perron–Frobenius eigenvalue is r_i as claimed. Let s_i denote the maximum value of $|\lambda|$ as λ varies over all the other eigenvalues of A_i'. Then $s_i < r_i$. We then obtain by linear algebra the following dichotomy, where

$$(4.4.24) \qquad \varepsilon_m = \sup_{F_i^m K} |u(x) - u(q_i)|.$$

LEMMA 4.4.1 *If u is harmonic and $\partial_n u(q_i) \neq 0$, then*

$$(4.4.25) \qquad c_1 r_i^m \leq \varepsilon_m \leq c_2 r_i^m,$$

while if $\partial_n u(q_i) = 0$ then

$$(4.4.26) \qquad \varepsilon_m \leq c(\delta)(s_i + \delta)^m \quad \text{for any } \delta > 0.$$

EXERCISES

4.4.1. Show that there exist values α and β such that

$$R(x, y) \leq c_1 |x - y|^\alpha \quad \text{and} \quad R(x, y) \geq c_2 |x - y|^\beta.$$

4.4.2. Rewrite the definition (4.4.8) directly in terms of u (rather than $u \circ F_w$).

4.4.3.* Compute $g(z, z')$ explicitly for SG$_3$.

4.4.4.* Compute $g(z, z')$ explicitly for the hexagasket with $N_0 = 3$.

4.4.5.* Show that (4.4.22) vanishes by using Exercise 4.2.4.

4.4.6.* Show that it is impossible to have $\partial_n h(q_i) = 0$ for all harmonic functions h.

4.4.7. Show that for harmonic functions the estimate (4.4.26) can be improved to
$$\varepsilon_m \leq c m^k s_i^m \text{ for some } k.$$

4.4.8. Show that

$$\Delta_\mu u(x) - \left(\int \psi_x^{(u)} d\mu \right)^{-1} \Delta_m u(x)$$

$$= \left(\int \psi_x^{(m)} d\mu \right)^{-1} \int \psi_x^{(m)}(y)(\Delta_\mu u(x) - \Delta_\mu u(y)) d\mu.$$

4.4.9.* On the hexagasket with $N_0 = 3$, define a tangential derivative at q_i by

$$\partial_T u(q_i) = \lim_{m \to \infty} 7^m (u(F_i^m q_{i+1}) - u(F_i^m q_{i-1}))$$

if the limit exists. (Here $7 = s_i^{-1}$.) Show that $\partial_T u(q_0)$ exists if u is harmonic, and the values $(u(q_i), \partial_n u(q_i), \partial_T u(q_i))$ uniquely determine the harmonic function. Show that if $u \in \text{dom } \Delta_\mu$ for μ the self-similar measure with equal weights, then $\partial_T u(q_i)$ exists. (Note that this is a stronger statement than Theorem 2.7.8 since we do not need to assume that $\Delta_\mu u$ satisfies a Hölder condition. The reason is that $r_i \mu_i = \frac{3}{7} \cdot \frac{1}{6} = \frac{1}{14} < s_i = \frac{1}{7}$ in this case.)

4.4.10.* On SG$_3$ define a tangential derivative at q_i by

$$\partial_T u(q_i) = \lim_{m \to \infty} (15)^m (u(F_i^m q_{i+1}) - u(F_i^m q_{i-1}))$$

if the limit exists. (Here $15 = s_i^{-1}$.) Show that $\partial_T u(q_i)$ exists if u is harmonic, and the values $(u(q_i), \partial_n u(q_i), \partial_T u(q_i))$ uniquely determine the harmonic function. Let μ be the self-similar measure with equal weights. Show that if $u \in \text{dom } \Delta_\mu$ and $\Delta_\mu u$ satisfies a Hölder condition of sufficiently high order, then $\partial_T u(q_i)$ exists. (Note that this is a weaker statement than Theorem 2.7.8 since we must prescribe the order of smoothness. The reason is that $r_i \mu_i = \frac{7}{15} \cdot \frac{1}{6} = \frac{7}{90} > s_i = \frac{1}{15}$.)

4.5 GEOGRAPHY IS DESTINY

SG is very far from being a homogeneous space. Nevertheless, one can easily be lulled into thinking that it is locally homogeneous, since given any two points x and y not on the boundary, there exist neighborhoods N_x and N_y that are isomorphic (but the isomorphism need not take x to y). Another related fact is that any pair of m-cells, except those touching the boundary, have isomorphic neighborhoods. But this is not true for general pcf fractals. For example, in the hexagasket there are some cells that have only two boundary points that are junction points, and some cells in which all three boundary points are junction points.

When we examine the behavior of functions on these fractals, the local differences become very striking. This is even true on SG, although the result is more subtle. We will look at the situation for harmonic functions only. Recall that $N_0 = \#V_0$, and the space $\mathcal{H}_0$ of harmonic functions has dimension N_0, each harmonic function being determined by its values on V_0. If h is harmonic, then so is $h \circ F_w$ for any word w, and the transformation $h \to f \circ F_w$ can be represented by an $N_0 \times N_0$ matrix A_w so that

$$(4.5.1) \qquad\qquad h \circ F_w \big|_{V_0} = A_w h \big|_{V_0}.$$

In fact we have

$$(4.5.2) \qquad\qquad A_w = A_{w_m} \cdot A_{w_{m-1}} \cdots A_{w_1},$$

where A_i are the harmonic extension matrices. If some of these matrices are not invertible, then the mapping is not onto. We only see a subspace of all possible harmonic functions when we zoom in on the cell $F_w K$, and the particular subspace depends on the particular cell. Another way of looking at it is that if x is a boundary point of $F_w K$ that is not a junction point, then $\partial_n h(x) = 0$. So the subspace must be contained in the space of all harmonic functions with vanishing normal derivative at each boundary point that is not a junction point. However, that is not the whole story. On the hexagasket, consider a cell $F_w K$ contained in a larger cell $F_{w'} K$ (so w' is a prefix of w). Suppose the boundary $F_{w'} V_0$ contains a nonjunction point. Then the space of harmonic functions restricted to $F_{w'} K$ is at most two-dimensional, so the same is true for $F_w K$. There are many examples where all three boundary points of $F_w K$ are junction points, so there are no vanishing normal derivative conditions to help describe the two-dimensional space here (as far as we know, it is never one-dimensional).

Thus we see that the space of restrictions to a cell $F_w K$ of all harmonic functions depends on the cell in a rather complicated way. So we say "geography is destiny": Where you are determines to some extent what kind of functions you see. To some extent there is a converse: Knowing a single function $h \circ F_w$ might help you determine where you are. Of course you can always see constants, so you would have to assume that $h \circ F_w$ is not constant (recall that $h \circ F_w$ may be constant even when h is not constant). The question then becomes: If you know the two-dimensional space of restrictions of harmonic functions, does that determine the cell? (This leaves out the relatively few cells that have three-dimensional spaces of restrictions.) This question has not been explored.

But there is a more subtle issue, and this already shows up in SG. Instead of asking what harmonic functions are possible, we should ask which are plausible. On SG, all the matrices A_i are invertible, so if your goal is to find $h \circ F_w = h'$ for any fixed h', you can find h, but it may be very difficult, since the matrix A_w^{-1} is very ill conditioned. If the cell is small, then we expect $h \circ F_w$ to be close to a constant. If we subtract off the constant, say $h \circ F_w - h(F_w q_0)$, we will get a function close to zero, so we want to multiply by a large constant c to get a normalized harmonic function, say

$$(4.5.3) \qquad\qquad c(h \circ F_w - h(F_w q_0)),$$

that satisfies the normalization condition

$$(4.5.4) \qquad \sum_i |f(q_i)|^2 = 1.$$

You therefore expect to be able to see any function satisfying $f(q_0) = 0$ and (4.5.4), but instead, you mainly see just one function. This is a consequence of the theory of products of random matrices. In fact, the product A_w may be thought of as a product of length $|w|$ selected from the set of matrices $\{A_i\}$ with equal probabilities, and chosen independently. The probability associated with each such product is $\mu(F_w K) = 3^{-m}$ for $m = |w|$. In fact, we want to work with the 2×2 matrices $\{\tilde{A}_i\}$ described in Exercise 1.3.4 in which we replace $\mathcal{H}_0$ by its quotient modulo constants. Then the matrices $\tilde{A}_w$ will be close to having rank 1 for most choices of w. There will always be a small number of cells for which we can't say anything, but for the preponderance of cells of level m, for moderately large values of m, there will be a harmonic function h_w associated to the cell $F_w K$ such that the normalized function (4.5.3) is close to $\pm h_w$. We could also say that $h \circ F_w$ may be written $c_1 + c_2 h_w$ up to a small error, for most harmonic functions h. We will not attempt to make these statements quantitative.

EXERCISES

4.5.1. Choose a normalized harmonic function on SG at random by choosing the boundary values at random on the 2-sphere defined by (4.5.4). Show that with probability 1 the harmonic function is almost symmetric in small neighborhoods of q_0, namely

$$\lim_{m \to \infty} \frac{h(F_0^m q_1) - h(F_0^m q_2)}{h(F_0^m q_1)} = 0.$$

4.5.2. The hexagasket has 6^m cells of level m. Estimate from above the number a_m of these cells $F_w K$ such that all cells $F_{w'} K$ containing $F_w K$ have all three boundary points being junction points. Show $\lim_{m \to \infty} a_m / 6^m = 0$.

4.6 NON-SELF-SIMILAR FRACTALS

The pcf fractals introduced in Section 4.1 possess two very strong properties: self-similarity and finite ramification. It is natural to wonder whether we can construct a Laplacian on fractals lacking these properties. In this section we look at some examples of fractals where we drop the self-similarity and replace it by a hierarchical structure. These fractals are built, level by level, from coarser to finer, by a finite set of construction principles. They also retain the finite ramification property. These fractals were introduced by Hambly in order to model "random fractals," as he used a random process to select the parameters in the construction. However, it is not necessary to do this, since there are many interesting results that hold for all values of the parameters.

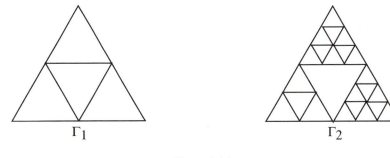

Figure 4.6.1

Figure 4.6.2

These fractals will again be limits of graphs $\{\Gamma_m\}$, with vertex sets increasing:

$$(4.6.1) \qquad V_0 \subseteq V_1 \subseteq V_2 \subseteq \cdots, \qquad V_* = \bigcup_m V_m,$$

and with V_* dense in K in a suitable sense. Γ_0 will be a complete graph with three vertices. At level m, V_m (and hence K) will be partitioned into m-cells, each with three boundary points in V_m, and the edge relation for Γ_m is that $x \underset{m}{\sim} y$ if and only if there is an m-cell containing x and y. To pass from Γ_m to Γ_{m+1} we assign a number $n(C)$, either 2 or 3, to each m-cell. If $n(C) = 2$ then we subdivide C into three $(m+1)$-cells, as in the construction of SG, while if $n(C) = 3$ we subdivide C into six $(m+1)$-cells, as in the construction of SG_3. Figure 4.6.1 shows one example (up to $m = 2$). Using the Euclidean metric we may define K to be the closure of V_*. We will call K a *hierarchical gasket*. If $n(C)$ only depends on the level m, we call K a *homogeneous hierarchical gasket*. Figure 4.6.2 shows one example. We may still index m-cells by words w of length m, where each w_i takes values in either $\{0, 1, 2\}$ or $\{0, 1, 2, 3, 4, 5\}$ depending on the value of $n(C_{(w_1, \ldots, w_{i-1})})$, but there are no mappings associated to cells.

To each cell we associate a resistance renormalization factor r_w defined inductively as $r_w = \frac{3}{5} r_{w'}$ if $n(C_w) = 2$ or $r_w = \frac{7}{15} r_{w'}$ if $n(C_w) = 3$, where $w = (w_1, \ldots, w_m)$ and $w' = (w_1, \ldots, w_{m-1})$. In other words,

$$(4.6.2) \qquad r_w = \left(\frac{3}{5}\right)^{m_2} \left(\frac{7}{15}\right)^{m_3},$$

where m_2 is the number of occurrences of 2 among $n(C_{(w_1,...,w_i)})$ for $1 \leq i \leq m$ and m_3 is the number of occurrences of 3. We assign conductances r_w^{-1} to all the edges in the cell C_w and then define the energy

$$(4.6.3) \qquad \mathcal{E}_m(u) = \sum_{|w|=m} \sum_{x,y \in C_w} r_w^{-1} |u(x) - u(y)|^2.$$

It is easy to see that if u is defined on V_{m-1}, then the extension to V_m that minimizes $\mathcal{E}_m$ is given by the harmonic extension algorithm for SG or SG$_3$ on each $(m-1)$-cell according to whether $n(C) = 2$ or 3, and moreover the harmonic extension $\tilde{u}$ satisfies $\mathcal{E}_m(\tilde{u}) = \mathcal{E}_{m-1}(u)$. Thus $\{\mathcal{E}_m(u)\}$ is increasing, and

$$(4.6.4) \qquad \mathcal{E}(u) = \lim_{m \to \infty} \mathcal{E}_m(u)$$

is well defined. All the properties of energy and harmonic functions discussed in Section 4.4 continue to hold, except for self-similarity. In particular, the diameter of a cell C_w in the resistance metric is on the order of r_w.

To define a Laplacian we could use any regular measure, but for simplicity we consider the natural one, which assigns

$$(4.6.5) \qquad \mu(C_w) = \mu_w = \left(\frac{1}{3}\right)^{m_2} \left(\frac{1}{6}\right)^{m_3}.$$

The discrete Laplacian Δ_m is defined by (4.4.4), where now $c_m(x, y) = r_w^{-1}$ if $x, y \in C_w$. The continuous Laplacian is defined by the weak formulation, which is equivalent to (4.4.5). In this case we can compute $\int \psi_x^{(m)} d\mu$ exactly. If $x \in V_m \setminus V_{m-1}$ and $x \in C_w$ then

$$(4.6.6) \qquad \int \psi_x^{(m)} d\mu = \begin{cases} \frac{2}{3}\mu_w & \text{if } n(C_w) = 2, \\ \begin{cases} \mu_w \\ \frac{2}{3}\mu_w \end{cases} & \text{if } n(C_w) = 3 \quad \text{and} \quad \begin{cases} \text{order}(x) = 3, \\ \text{order}(x) = 2. \end{cases} \end{cases}$$

If $x \in V_{m-1}$ let $\{C_{w^{(j)}}\}$ denote the m-cells containing x (there may be one, two, or three such). Then

$$(4.6.7) \qquad \int \psi_x^{(m)} d\mu = \sum_j \frac{1}{3}\mu_{w^{(j)}}.$$

We leave the verification to the exercises.

The definition of normal derivative may be given as follows. Suppose $x \in V_m$ and C is an m-cell containing x. Then there is a sequence $w^{(j)}$ with $|w^{(j)}| = j$ (for $j \geq m$) with $x \in C_{w^{(j)}} \subseteq C$. Then

$$(4.6.8) \qquad \partial_n^C u(x) = \lim_{j \to \infty} (r_{w^{(j)}})^{-1} \sum_{y \in \partial C_{w^{(j)}}} (u(x) - u(y)).$$

Note that there are always exactly two terms in the summation. As before, these derivatives exist for $u \in \text{dom } \Delta_\mu$, the localized Gauss–Green formulas hold, and the matching conditions are

$$(4.6.9) \qquad \sum_C \partial_n^C u(x) = 0,$$

where the sum extends over all m-cells containing x (two or three, for a non-boundary point).

The construction of the Green's function may be done by combining the constructions of SG and SG_3 in the appropriate way. Once again we want to define G by (4.4.11) in such a way that (4.4.14) holds, but we need to modify the definition of G_M to read

$$(4.6.10) \quad G_M(x, y) = \sum_{m=0}^{M} \sum_{|w|=m} \sum_{z,z' \in (V_{m+1} \setminus V_m) \cap C_w} r_w g(z, z') \psi_z^{(m+1)}(x) \psi_{z'}^{(m+1)}(y),$$

where $g(z, z')$ is as in SG if $n(C_w) = 2$ or as in SG_3 if $n(C_w) = 3$ (Exercise 4.3.3). We leave it to the exercises to verify that (4.4.14) holds.

EXERCISES

4.6.1. Show that every cell in a hierarchical gasket is itself a hierarchical gasket.

4.6.2. Show that the diameter of a cell C_w in the resistance metric is bounded above and below by multiples of r_w. Use this to conclude that the resistance metric defines the same topology as the Euclidean metric.

4.6.3. Verify (4.6.5) and (4.6.6).

4.6.4.* Verify that if G_M is defined by (4.6.9), then (4.4.14) holds.

4.6.5. Show that the D_3 symmetry group acts isomorphically on any homogeneous hierarchical gasket. Are there any nonhomogenous examples where the same is true?

4.7 NOTES AND REFERENCES

The definition of pcf fractals and the construction of energy and Laplacian given a solution to the renormalization problem is from [Kigami 1993]. He uses a broader definition. In particular, he does not require that the fractal embed in any Euclidean space. The requirement that the boundary points be fixed points of the mappings does eliminate some interesting examples, such as the Hata tree.

There is an extensive literature on the renormalization problem, beginning with [Hattori et al. 1987]. [Lindstrøm 1990] introduces a class of fractals called nested fractals (the Lindstrøm snowflake is but one example); roughly spseaking, these are the pcf fractals with full dihedral symmetry. It is proved there by nonconstructive means (the Brouwer fixed point theorem) that solutions exist for all nested fractals. [Sabot 1997] proves that the solution is unique, and [Metz 1996] proves that it is attracting. This means the solution is constructive, but still there is no explicit formula for the snowflake. General methods for solving the renormalization problem, and many specific examples, may be found in [Hambly et al. *], [Metz 1993, 1995, 1997, 2003a, 2003b, 2004, 2005], [Peirone 2000], and [Peirone *]. In particular, Exercise 4.3.4 showing the failure of uniqueness is from [Metz 1993]. [Peirone 2000] shows that if the renormalization problem has a unique solution, then it is attracting. [Sabot 1997] explicitly describes all self-similar energies on SG. The general case turns out to be much more involved than the examples with bilateral symmetry worked out in Section 4.3. However, if you consider a different i.f.s. that generates SG, obtained by adding twists to the mappings, then it becomes possible to give a simple description of all self-similar energies, as shown in [Cucuringu & Strichartz *]. A somewhat incomplete description of all energies (without self-similarity) is given in [Meyers et al. 2004].

The definition of tangential derivatives on pcf fractals with D_3 symmetry and Exercises 4.4.9 and 4.4.10 are from [Strichartz 2000a], where some results on local behavior of harmonic functions and functions in the domain of the Laplacian may be found (see also [Teplyaev 2000]).

The ideas in Section 4.5 are discusssed in [Öberg et al. 2002]. Some extensions to functions in the domain of the Laplacian may be found in [Pelander & Teplyaev *]. For the theory of products of random matrices, see [Bougerol 1985].

The examples in Section 4.6 are introduced in [Hambly 1992, 1997]. Many properties of these hierarchical gaskets are known, including heat kernel estimates (see [Barlow & Hambly 1997], [Hambly 2000]). Of course, it is not necessary to restrict attention to just the SG and SG_3 constructions; one can allow analogous subdivision schemes SG_k by subdividing each side of the triangle into k segments. The homogeneous hierarchical fractals satisfy a version of spectral decimation, as is shown in [Drenning & Strichartz *].

A more general theory of energy and Laplacian on fractals that can be realized as limits of electric networks is given in [Kigami 2003].

The existence of localized eigenfunctions on pcf fractals with symmetry is discussed in [Barlow & Kigami 1997]. It is also possible to extend the method of spectral decimation to a class of pcf fractals, as shown in [Shima 1996].

Chapter Five

Further Topics

5.1 POLYNOMIALS, SPLINES, AND POWER SERIES

On the unit interval, the space of polynomials may be characterized as the space of solutions to $\Delta^k u = 0$ for some k. So it is natural to define a *polynomial* on any of the fractals K discussed in Chapter 4 as a solution of the same equation. In this section we will restrict attention to the case $K = SG$ with the standard Laplacian. We define $\mathcal{H}_k = \{u : \Delta^{k+1} u = 0\}$ and refer to these functions as *multiharmonic* (or *biharmonic* for $k = 1$, *triharmonic* for $k = 2$, etc.). We have seen that $\mathcal{H}_0$, the space of harmonic functions, is three-dimensional, and a function in $\mathcal{H}_0$ is uniquely determined by its boundary values $\{h(q_i)\}$.

If we want a function in $\mathcal{H}_1$, then it satisfies the equation $\Delta u = h$ with $\Delta h = 0$, so h is harmonic. Then

$$(5.1.1) \qquad u(x) = \int_K G(x, y)h(y)d\mu(y)$$

gives us a solution. But by the definition of the Green's function, u must vanish at the boundary points. It is easy to see that

$$(5.1.2) \qquad u(x) = \int_K G(x, y)h(y)d\mu(y) + h_1(x),$$

where h and h_1 are harmonic, gives the general function in $\mathcal{H}_1$, so $\mathcal{H}_1$ is six-dimensional. A function in $\mathcal{H}_1$ is uniquely determined by the boundary values of u and Δu, and we may give an easy basis for $\mathcal{H}_1$ by prescribing one of these values to be 1 and the other five to be 0. Also, by using properties of the Green's function, we can give an explicit biharmonic extension algorithm to find the values of u on V_1, and then recursively on V_m for all m.

But if we think of the situation on the interval we realize that this basis is not at all natural or useful: It corresponds to describing a cubic polynomial by the boundary values of the polynomial and its second derivative. Why did we skip the first derivative? It is an easy exercise to show that either way, $(p(0), p(1), p''(0), p''(1))$ or $(p(0), p(1), p'(0), p'(1))$, will work, but clearly the second choice is better. In the case of SG the natural data to give are the boundary values of the function and its normal derivative, $\{u(q_i), \partial_n u(q_i)\}$. These six values uniquely determine $u \in \mathcal{H}_1$, as may be seen by explicitly solving the equations (see the exercises). We may then define a better basis by requiring

$$(5.1.3) \qquad \begin{cases} f_{0i}(q_j) = \delta_{ij}, & \partial_n f_{0i}(q_j) = 0, \\ f_{1i}(q_j) = 0, & \partial_n f_{1i}(q_j) = \delta_{ij}. \end{cases}$$

We have already seen that the space of piecewise harmonic splines plays the same role as piecewise linear functions on the interval. But piecewise linear functions must be quite bumpy, and for many purposes they don't give acceptable approximations to functions that are supposed to be smooth. By using piecewise cubic polynomials we may match up the derivatives at the knots to produce a C^1 function. The analogous idea on SG is to use piecewise biharmonic splines and enforce the matching conditions at the junction points. Recall that we defined the space $S(\mathcal{H}_0, V_m)$ of piecewise harmonic splines (Definition 1.4.3) to be the continuous functions that are harmonic on each m-cell. The continuity is a matching condition on the boundary values of $u \circ F_w$ ($|w| = m$) at junction points. Similarly we may define $S(\mathcal{H}_1, V_m)$ to be the space of functions u such that $u \circ F_w \in \mathcal{H}_1$ for $|w| = m$ (this is the same as saying $u|_{F_w K}$ agrees with a function in $\mathcal{H}_1$) such that for each junction point $x = F_w q_i = F_{w'} q_{i'}$ we have two matching conditions: The values are equal, $u \circ F_w(q_i) = u \circ F_{w'}(q_{i'})$, and the normal derivatives sum to zero, $\partial_n(u \circ F_w)(q_i) + \partial_n(u \circ F_{w'})(q_{i'}) = 0$. Note that this implies $u \in \mathrm{dom}_{L^2} \Delta$. We can't assert that $u \in \mathrm{dom}\,\Delta$ because Δu need not be continuous. It is easy to see that $S(\mathcal{H}_1, V_m)$ has dimension $2(\#V_m)$, and we can form a basis by gluing localized versions of the functions defined in (5.1.3). These spline spaces may be used in standard finite element method approximations to solutions to differential equations involving the Laplacian. By doubling the dimension in comparison with $S(\mathcal{H}_0, V_m)$ we obtain vastly better approximation rates.

In a similar way we may define higher order spline spaces $S(\mathcal{H}_k, V_m)$. We impose matching conditions at junction points as follows:

$$(5.1.4) \qquad \Delta^j(u \circ F_w)(q_i) = \Delta^j(u \circ F_{w'})(q_{i'}) \quad \text{for } j \le \frac{k}{2},$$

$$(5.1.5) \qquad \partial_n \Delta^j(u \circ F_w)(q_i) + \partial_n \Delta^j(u \circ F_{w'})(q_{i'}) = 0 \quad \text{for } j < \frac{k}{2}.$$

Again, it is necessary to find a better basis for $\mathcal{H}_k$ based on boundary values of $\Delta^j u$ and $\partial_n \Delta^j u$ with j as above. This is possible, but it is quite complicated to describe explicitly. It is also possible to define spline spaces based on irregular partitions.

The basis for $\mathcal{H}_1$ described in (5.1.3) treats all three boundary points on an equal footing. But for some applications we need to focus on a single boundary point, say q_0. In particular, we would like to examine the analog of Taylor polynomials to characterize the local behavior of a function in a neighborhood of a point. In Section 2.7 we introduced the notion of the 1-jet of a function u at q_0, namely the triple $(u(q_0), \partial_n u(q_0), \partial_T u(q_0))$. For this to make sense we had to assume that $u \in \mathrm{dom}\,\Delta$ and Δu is Hölder continuous (of any positive order). We also found a basis for the harmonic functions, which we now rename as follows:

$$(5.1.6) \qquad P_{01} \equiv 1, \qquad P_{02} = -\frac{1}{2}(h_1 + h_2), \qquad P_{03} = \frac{1}{2}(h_1 - h_2),$$

having jets $(1, 0, 0)$, $(0, 1, 0)$, and $(0, 0, 1)$. In a similar way we may define the k-jet of u at q_0 to consist of the $3k$-tuple of values $(\Delta^j u(q_0), \partial_n \Delta^j u(q_0), \partial_T \Delta^j u(q_0))$ for $j < k$. The k-jet is well defined if $u \in \mathrm{dom}\,\Delta^k$ and $\Delta^k u$ is Hölder continuous. Then we want to define polynomials $P_{ki} \in \mathcal{H}_k$ to have k-jet consisting of all 0s except for one 1:

(5.1.7)
$$\begin{cases} \Delta^j P_{ki}(q_0) = \delta_{jk}\delta_{i1}, \\ \partial_n \Delta^j P_{ki}(q_0) = \delta_{jk}\delta_{i2}, \\ \partial_T \Delta^j P_{ki}(q_0) = \delta_{jk}\delta_{i3} \end{cases}$$

for $j \le k$. Note that this means

(5.1.8) $\Delta P_{ki} = P_{(k-1)i}$,

and we can use this to recursively determine P_{ki}, since then

(5.1.9) $P_{ki}(x) = -\displaystyle\int G(x,y) P_{(k-1)i}(y) d\mu(y) + h(x)$

for a harmonic function $h(x)$ that may be determined by the $j = 0$ case of (5.1.7).

The polynomials $\{P_{ji}\}$ for $j < k$ form a basis for $\mathcal{H}_{k-1}$. If $u \in$ dom Δ^k and $\Delta^k u$ is Hölder continuous, then we can use this basis to create a function in $\mathcal{H}_{k-1}$ with the same k-jet as u at q_0 in the obvious way. This is the analog of the Taylor polynomial of order $2k - 1$ on the interval. We can then prove order of approximation results analogous to (2.7.22) and (2.7.23) (the $k = 1$ case) with 5^{-m} replaced by 5^{-km}. In other words, the higher we take k the faster the Taylor polynomial approximates the function near q_0.

We can also consider the analog of power series (about the point q_0) by letting $k \to \infty$. The functions $\{P_{ki}\}$ are analogous to the functions $\left\{ \frac{x^j}{j!} \right\}$, so the key idea is to estimate the size of $\| P_{ki} \|_\infty$. Presumably the supremum is attained at the boundary (q_1 or q_2), but we only have experimental evidence for this. It turns out that the decay of $\| P_{ki} \|_\infty$ for $i = 1$ or 3 is superexponential (faster than $c_r r^{-k}$ for any r), but $\| P_{k2} \|_\infty$ is exactly exponential, namely $O(\lambda_2^{-k})$, where $\lambda_2 = 135.572126\ldots$ is the second nonzero Neumann eigenvalue (belonging to the 6-series with $m_0 = 1$). This fact has a number of interesting consequences. The positive result is that any power series

(5.1.10) $\displaystyle\sum_{k=0}^{\infty} \sum_{i=1}^{3} c_{ki} P_{ki}(x)$

converges uniformly if the coefficients satisfy an estimate of the form

(5.1.11) $|c_{ki}| \le M r^k$ for some $r < \lambda_2$.

We will call functions of this form *entire analytic*. It can be shown that such power series may be "rearranged" around either of the other two boundary points q_1 or q_2, and the coefficients again satisfy (5.1.11). It is possible to obtain convergence of (5.1.10) under weaker hypotheses than (5.1.11) for $i = 1, 3$, but then it does not follow that those power series may be rearranged, which explains why we choose to define analytic functions in such a way that not all convergent power series represent analytic functions.

It also follows that not all eigenfunctions of the Laplacian are entire analytic functions. The general λ-eigenfunction power series looks like

(5.1.12) $\displaystyle\sum_{k=0}^{\infty} (-\lambda)^k (a_1 P_{k1}(x) + a_2 P_{k2}(x) + a_3 P_{k3}(x))$, and this yields a three-dimensional space of entire analytic functions if $|\lambda| < \lambda_2$. But for $\lambda = \lambda_2$, the entire eigenspace, which is three-dimensional because λ_2 is not a Dirichlet eigenvalue, consists of Neumann eigenfunctions, which would force $a_2 = 0$. This explains why (5.1.12) must diverge for $\lambda = \lambda_2$ and $a_2 \neq 0$, and hence why the estimates for $\|P_{k2}\|_\infty$ are best possible. Of course, there are also eigenspaces of dimension greater than three, so (5.1.12) would not be adequate to give all of those eigenfunctions.

There is a really striking difference between polynomials on SG and polynomials on an interval when it comes to the global theory: The analog of the Weierstrass approximation theorem is false. In fact, polynomials on SG have some rather strong global properties that easily show that they cannot uniformly approximate all continuous functions. For example, let u be any joint D–N eigenfunction. Then

$$(5.1.13) \qquad \int P(x)u(x)d\mu(x) = 0$$

follows easily by repeated use of the Gauss–Green formula, with no boundary terms arising because $u(q_i)$ and $\partial_n u(q_i)$ both vanish. It follows that any uniform (or even L^2) limit of polynomials also satisfies (5.1.13). In fact, the converse statement is also true: Any function $P(x)$ satisfying (5.1.13) for all joint D–N eigenfunctions is an L^2 limit of polynomials. Another type of identity satisfied by all polynomials, and hence all limits of polynomials, is

$$(5.1.14) \qquad P(x) + P(R_1 x) + P(R_2 x) = P(\rho_0 x) + P(\rho_1 x) + P(\rho_2 x),$$

where ρ_i are the reflections and R_i are the rotations in the dihedral-3 symmetry group of SG and x is any point. There are also local versions obtained by replacing P by $P \circ F_w$ in (5.1.14). It also follows from (5.1.14) that for any polynomial, the sum of the tangential derivatives at the boundary points vanishes (and similarly at the boundary points of any cell). This fact does not extend to uniform limits of polynomials, however, since the tangential derivatives might not exist.

EXERCISES

5.1.1. Find an explicit formula for a cubic polynomial given the values $(p(0), p(1), p''(0), p''(1))$. Do the same in terms of the values $(p(0), p(1), p'(0), p'(1))$.

5.1.2.* Give an explicit description of the basis for $\mathcal{H}_1$ determined by (5.1.3).

5.1.3. Show that $u \circ F_w \in \mathcal{H}_k$ if and only if there exists $v \in \mathcal{H}_k$ such that $u = v$ on $F_w K$.

5.1.4. Show by induction that there is a unique solution to (5.1.7) in $\mathcal{H}_k$.

5.1.5. Prove that (5.1.13) holds for polynomials.

5.1.6. Let u be a Dirichlet eigenfunction. Show that (5.1.13) holds for all polynomials if and only if u is also a Neumann eigenfunction.

5.1.7. Show that (5.1.14) implies

$$\sum_{i=0}^{2} \partial_T P(q_i) = 0$$

for any polynomial P.

5.1.8. Prove the scaling identities $P_{k1} \circ F_0^m = \left(\frac{1}{5^k}\right)^m P_{k1}$, $P_{k2} \circ F_0^m = \left(\frac{3}{5^{k+1}}\right)^m P_{k2}$, and $P_{k3} \circ F_0^m = \left(\frac{1}{5^{k+1}}\right)^m P_{k3}$.

5.2 LOCAL SYMMETRIES

The interval has only two isometries, the identity and the flip $x \to (1-x)$. But surely, on an informal level, there is more symmetry to the interval. One way to make this apparent is to allow discontinuous isometries. For example, split the interval in half and flip both halves. This mapping,

$$(5.2.1) \qquad\qquad F(x) = \begin{cases} \frac{1}{2} - x, & \text{if } 0 \leq x < \frac{1}{2}, \\ \frac{3}{2} - x, & \text{if } \frac{1}{2} < x \leq 1, \end{cases}$$

while not well defined at $x = \frac{1}{2}$, is a local isometry everywhere else. More to the point, if u is a continuous function satisfying $u(0) = u(1)$, then in fact $u \circ F$ is a well-defined continuous function satisfying the same condition. In particular, if k is even, then $\sin \pi k x$ is preserved by this mapping, and $\cos \pi k x$ is mapped to its negative. So at least some of the Dirichlet or Neumann eigenfunctions are symmetric under this symmetry. Similarly, we could cut the interval up into 2^m dyadic subintervals of length 2^{-m} and flip all of them. This type of reasoning can help explain the repetitive pattern we see in the trigonometric functions $\sin \pi k x$ and $\cos \pi k x$.

We may use similar reasoning on SG, but we obtain somewhat different conclusions. We consider an infinite sequence of "rip and flip" transformations that are local isomorphisms with a finite set of discontinuites. The first of these, S_1, simply reflects each of the three 1-cells $F_i K$ keeping q_i fixed. The points in $V_1 \setminus V_0$ are points of discontinuity for S_1, but if u is continuous and assumes the same value at these three points, then $u \circ S_1$ is well defined and continuous. In particular, if u is invariant under D_3, then it is invariant under S_1. Figure 5.2.1 gives a symbolic picture of S_1, with the dotted lines indicating the flip axes.

Figure 5.2.2 gives a similar symbolic picture of S_2, the second in our sequence of local symmetries. Note that the three boundary 2-cells are held fixed, while the other six 2-cells are flipped. Note that these are the cells that line the inner upside-down triangle $\tilde{T}$. There are now six points of discontinuity, three points in

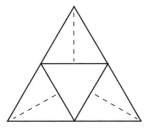

Figure 5.2.1

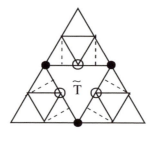

Figure 5.2.2

$V_1 \setminus V_0$, marked by solid dots, and three more points along $\tilde{T}$ marked by open dots in Figure 5.2.2. Note that S_2 interchanges these two types of points. So if u is continuous and assumes constant values on these two types of points, then $u \circ S_2$ is well defined and continuous. In particular, this is true if u is invariant under D_3.

We repeat this process to define successively $S_3, S_4, \ldots$. The map S_m will flip the $3 \cdot 2^{m-1}$ m-cells that line $\tilde{T}$ and leave all the other m-cells fixed. The points of discontinuity of S_m all lie along $\tilde{T}$ and consist of the points of discontinuity of S_{m-1} together with the fixed points of the flips in S_{m-1} that lie along $\tilde{T}$. In order for $u \circ S_m$ to be well defined and continuous, we have to impose more and more conditions on u as m increases. Surprisingly, there is a way to guarantee that all these conditions hold, and moreover show that u is invariant under all the maps S_m.

For example, suppose u is a Dirichlet eigenfunction in the 2-series. Then u is D_3 invariant, and the eigenspace has multiplicity 1. Since u is D_3 invariant it is clearly S_1 invariant, so $u \circ S_2$ is continuous. But we claim that in fact $u \circ S_2 = u$. Why? Because $u \circ S_2$ is also an eigenfunction (with the same eigenvalue). This is obvious at all points except perhaps the six points where S_2 is discontinuous. But at these points S_2 merely transposes the eigenvalue equation: The eigenvalue equation for $u \circ S_2$ at a solid dot point is identical to the eigenvalue equation for u at an open dot point. Since u and $u \circ S_2$ are identical on the outer cells and the eigenspace is three-dimensional, it follows that they are equal everywhere. We can then repeat this argument inductively to conclude that $u \circ S_m = u$ for all m.

We will say that u exhibits a *local symmetry* near $\tilde{T}$. What does this imply about u? For one thing, it implies that u is constant on $\tilde{T}$. Indeed, it is easy to show by induction that u is constant on the set of discontinuities of S_m. For example, the invariance $u \circ S_2 = u$ implies that u takes the same value at the solid dot points as it does at the open dot points, and so on. Since the set of discontinuities of S_m for all m is a dense subset of $\tilde{T}$, the continuity of u implies that it is constant on $\tilde{T}$.

But the local symmetry implies much more: It implies that the restriction of u to each of the $3 \cdot 2^{m-1}$ m-cells lining $\tilde{T}$ is the same. So we see a type of periodicity akin to that of the trigonometric functions, provided we zoom in to a neighborhood of $\tilde{T}$. But unlike the case of trigonometric functions, we see periodicity of arbitrarily high frequency in a single eigenfunction. Of course, we have to look very carefully to see it, since the function is converging to a constant and so shows only slight variation near $\tilde{T}$.

This local symmetry is not restricted to just these Dirichlet eigenfunctions. If we choose any eigenvalue not in the Dirichlet spectrum then the space of eigenvalues is three-dimensional, and if we prescribe the same value at all three boundary points, we obtain a D_3 invariant eigenfunction. We may then repeat the above reasoning to conclude that this eigenfunction has local symmetry. The same will be true for a biharmonic function that satisfies $\Delta u = 1$ and $u(q_i) = 0$ for all i.

There are also local symmetries about other subsets of SG, such as the line segment L joining q_1 and q_2. Let S_1' denote the discontinuous mapping that flips $F_1 K$ and $F_2 K$, keeping $F_1 q_0$ and $F_2 q_0$ fixed. If u is invariant under the reflection fixing q_0, then $u \circ S_1'$ is well defined. Similarly, S_m' will flip all the 2^m m-cells that line L and keep all the other m-cells unchanged.

Suppose λ is any eigenvalue with a three-dimensional eigenspace. We can then find an eigenfunction satisfying $u(q_1) = u(q_2) = u(F_1 q_2)$. Such an eigenfunction will have to be invariant under the reflection fixing q_0, and so will be invariant under $u \circ S_1'$. We can then run an induction argument to show that $u \circ S_m' = u$ for all m. So u is constant along L and exhibits a local symmetry near L. In fact, we can describe the behavior of u near L more explicitly (see the exercises).

Local symmetries may also be found on other very symmetric fractals, such as the polygaskets.

EXERCISES

5.2.1. Show that for a generic eigenvalue λ, the values $u(q_1), u(q_2), u(F_1 q_2)$ uniquely determine an eigenfunction. In particular, if $u(q_1) = u(q_2) = u(F_1 q_2) = 1$, show that $u \circ S_1' = u$, and then by induction $u \circ S_m' = u$.

5.2.2. Show that the eigenfunction in Exercise 5.2.1 takes on the value $1 - \frac{\lambda_m}{2}$ at the points $F_w q_0$ for $|w| = m$ with all $w_i = 1$ or 2 (in other words, the vertices of height 2^{-m} above L). Similarly, show that the D_3 invariant eigenfunctions taking the value $u = 1$ on $\tilde{T}$ take the value $1 - \frac{\lambda_m}{2}$ at the vertices at distance 2^{-m} from $\tilde{T}$.

5.2.3. Describe local symmetries on the pentagasket and hexagasket.

5.3 ENERGY MEASURES

At the end of Section 1.4 we indicated that energy may be regarded as the integral of a certain energy measure. This idea is valid on any pcf fractal with regular harmonic structure. For a function $u \in \mathrm{dom}\,\mathcal{E}$ we define $\nu_u(C)$ for any cell C by the same definition as $\mathcal{E}(u)$ except that the sums are restricted to edges lying in C. Aside from the strict positivity, this defines a regular measure on K using additivity. By the self-similarity of energy, we have

$$(5.3.1) \qquad \nu_u(F_w K) = r_w^{-1} \mathcal{E}(u \circ F_w).$$

Then energy is given by

(5.3.2)
$$\mathcal{E}(u) = \nu_u(K) = \int_K 1 d\nu_u.$$

Similarly, for any pair of functions $u, v \in \text{dom}\,\mathcal{E}$, we may define a signed measure $\nu_{u,v}$ so that

(5.3.3)
$$\mathcal{E}(u, v) = \nu_{u,v}(K) = \int_K 1 d\nu_{u,v},$$

and $\nu_{u,v}$ is a symmetric bilinear function of u and v with $\nu_u = \nu_{u,u}$. There is yet another formula for these measures that goes under the name *carré du champs*, namely

(5.3.4)
$$\int_K f d\nu_{u,v} = \frac{1}{2}\mathcal{E}(fu, v) + \frac{1}{2}\mathcal{E}(u, fv) - \frac{1}{2}\mathcal{E}(f, uv)$$

for any $f \in \text{dom}\,\mathcal{E}$ (see the exercises).

Choose an orthonormal basis $\{h_j\}$ for the harmonic functions modulo constants in the energy inner product. Then define the *Kusuoka measure* ν by

(5.3.5)
$$\nu = \sum_j \nu_{h_j}.$$

It is easy to see that ν is independent of the choice of orthonormal basis (this was Exercise 1.4.4 in the case $K = SG$). To indicate why this is a natural object we digress to look at the analogous construction on the unit disk D in the plane with energy

(5.3.6)
$$\mathcal{E}(u) = \int_D |\nabla u|^2 dx dy.$$

In this case there is an infinite orthonormal basis of harmonic functions that can be computed explicitly, and by interpreting (5.3.5) as an infinite series (yielding an infinite measure) we can show that ν is a multiple of the Riemannian measure associated with the hyperbolic metric on the disk. We leave the details of the explicit calculation to the exercises, but note that there is a very simple reason why this must be the case. It is well known that the energy (5.3.6) in any planar domain is invariant under conformal mappings (the distortion in $|\nabla u|^2$ exactly cancels the distortion in the area measure). Also, there is a transitive group of conformal mappings Φ on D (a class of Möbius transformations) that are isometries in the hyperbolic metric. In particular, the image under Φ of an orthonormal basis of harmonic functions is another orthonormal basis of harmonic functions (conformal maps preserve harmonic functions). Thus the measure ν is invariant under each mapping Φ, and the Riemannian measure is the unique, up to a constant multiple, invariant measure under the group.

In the fractal setting, the main results are that all the measures $\nu_{u,v}$ are asolutely continuous with respect to ν, whereas ν is singular with respect to any self-similar measure in most cases. The absolute continuity means, by the Radon–Nikodym theorem, that for every pair $u, v \in \text{dom}\,\mathcal{E}$ there exists a function $f_{u,v} \in L^1(dv)$ such that

$$(5.3.7) \qquad\qquad\qquad v_{u,v} = f_{u,v}dv.$$

The mapping $(u, v) \to f_{u,v}$ must be bilinear in u and v and continuous. Beyond that, not much is known about it.

THEOREM 5.3.1 *For any $u, v \in \text{dom}\,\mathcal{E}$, the measure $v_{u,v}$ is absolutely continuous with respect to v.*

Proof: For simplicity we give the proof for $u = v$ (use polarization for the general case). To show absolute continuity, we have to show that for any $\varepsilon > 0$ there exists $\delta > 0$ such that $v(A) \leq \delta$ implies $v_u(A) \leq \varepsilon$.

First we observe that this is very easy if u is piecewise harmonic. Indeed, on any cell C on which u is harmonic we may write $u = \sum a_j h_j$ (the coefficients vary with the cell, of course), and so $v_u = \sum\sum a_j a_k v_{h_j,h_k}$ on C. But it is clear that $|v_{h_j,h_k}(A)| \leq v(A)$ for any set A by a Cauchy–Schwartz argument. Thus there is a constant c such that

$$(5.3.8) \qquad\qquad\qquad v_u(A) \leq cv(A),$$

and this implies absolute continuity.

For the general case, we use the density of piecewise harmonic functions in $\text{dom}\,\mathcal{E}$ in energy norm. So given $u \in \text{dom}\,\mathcal{E}$ and $\varepsilon > 0$, we may write $u = u_1 + u_2$ with u_1 piecewise harmonic and $\mathcal{E}(u_2) \leq \varepsilon$. Then $v_u = v_{u_1} + v_{u_2} + 2v_{u_1,u_2}$. We can make $v_{u_2}(A)$ and $v_{u_1,u_2}(A)$ small for all A just using $\mathcal{E}(u_2) \leq \varepsilon$, and then use (5.3.8) to control $v_{u_2}(A)$ in terms of $v(A)$. This shows the absolute continuity. $\qquad\square$

THEOREM 5.3.2 *Let h be a nonconstant harmonic function with respect to the standard energy on SG. Then v_h is not absolutely continuous with respect to the standard measure.*

Proof: By the Radon–Nikodym theorem, we have to show that there does not exist a function $f \in L^1(d\mu)$ such that

$$(5.3.9) \qquad\qquad \mathcal{E}_C(h) = \int_C f d\mu \quad \text{for every cell } C.$$

So, suppose such a function exists, and let

$$(5.3.10) \qquad\qquad f_m = \sum_{|w|=m} \left(3^m \int_{F_w K} f d\mu \right) \chi_{F_w K}$$

be the piecewise constant approximation to f at level m. By standard measure theory we would have $f_m \to f$ in the L^1 norm. We will show that this is impossible.

Note that $f_m(x) = 3^m \mathcal{E}_{F_w K}(h)$ for $x \in F_w K$. Thus on $F_w F_i K$ the function f_m equals $3^m \mathcal{E}_{F_w K}(h)$, while f_{m+1} equals $3^{m+1} \mathcal{E}_{F_w F_i K}(h)$. It follows that

$$(5.3.11) \qquad \| f_m - f_{m+1} \|_1 = \sum_{|w|=m} \sum_{i=0}^{2} \left| \frac{1}{3} \mathcal{E}_{F_w K}(h) - \mathcal{E}_{F_w F_i K}(h) \right|,$$

and also

(5.3.12)
$$\| f_m \|_1 = \sum_{|w|=m} |\mathcal{E}_{F_w K}(h)|.$$

We claim there exists a positive constant $\varepsilon > 0$ such that

(5.3.13)
$$\sum_{i=0}^{2} \left| \frac{1}{3} \mathcal{E}_{F_w K}(h) - \mathcal{E}_{F_w F_i K}(h) \right| \geq \varepsilon |\mathcal{E}_{F_w K}(h)|.$$

Clearly (5.3.13) is independent of the level and is just a quantitative form of the statement that it is impossible to split the energy equally among all three cells $F_i K$ (Exercise 1.3.5). We leave the details to the exercises. Combining (5.3.13) with (5.3.11) and (5.3.12) yields

(5.3.14)
$$\| f_m - f_{m+1} \|_1 \geq \varepsilon \| f_m \|_1,$$

and this contradicts $f_m \to f \neq 0$ in L^1. $\qquad\qquad\square$

Notice the contrast between SG and I; for any nonconstant harmonic function on I, the energy on any interval splits exactly in half when you split the interval in half. Clearly the proof of the above theorem could be extended to other fractals where we have enough explicit information on the harmonic extension algorithm. On the other hand, with a bit more work, it can be shown that ν_h is actually singular with respect to μ (or any other self-similar measure). In fact, we can more or less explain the singularity qualitatively by the fact that ν_h (and hence ν) tends to put more heavy weight on cells containing points in V_m for small values of m.

We can use Theorem 5.3.2 together with (5.3.4) to give yet another proof of Corollary 2.7.5 that a nonconstant $u \in \text{dom} \, \Delta$ does not permit $u^2 \in \text{dom} \, \Delta$. Actually we need the extension of Theorem 5.3.2 to nonconstant functions in $\text{dom} \, \mathcal{E}$, which follows by the reasoning in the proof of Theorem 5.3.1. We let $v \in \text{dom}_0 \mathcal{E}$ and observe that if $\Delta(u^2)$ were to exist we would have

(5.3.15)
$$- \int \left(\Delta u^2 \right) v \, d\mu = \mathcal{E} \left(u^2, v \right).$$

But in fact (5.3.4) shows that

(5.3.16)
$$\mathcal{E} \left(u^2, v \right) = 2 \mathcal{E}(u, uv) - 2 \int v \, d\nu_u.$$

Also $uv \in \text{dom}_0 \mathcal{E}$ so

(5.3.17)
$$\mathcal{E}(u, uv) = - \int (\Delta u) uv \, d\mu.$$

Combining (5.3.16) and (5.3.17) we see that

(5.3.18)
$$\mathcal{E}(u^2, v) = -2 \int (u \Delta u) v \, d\mu - 2 \int v \, d\mu_u.$$

We may interpret (5.3.18) in light of Exercise 2.1.5 as saying that $u^2 \in \text{dom}_{\mathcal{M}} \Delta$ with

(5.3.19)
$$\Delta(u^2) = 2u \Delta u \, d\mu + 2 d\nu_u,$$

but the right-hand side of (5.3.19) is not absolutely continuous with respect to μ. Note that (5.3.19) is the analog of the familiar identity

$$(5.3.20) \qquad \Delta(u^2) = 2u \cdot \Delta u + 2\nabla u \cdot \nabla u$$

on Euclidean space or, more generally, Riemannian manifolds.

In fact, the same reasoning shows that if $u \in \operatorname{dom} \Delta_\nu$ then $u^2 \in \operatorname{dom} \Delta_\nu$ with

$$(5.3.21) \qquad \Delta_\nu(u^2) = 2u \Delta_\nu u + 2\frac{d\nu_u}{d\nu}.$$

In this respect, Δ_ν is better behaved than the standard Laplacian.

EXERCISES

5.3.1. Prove (5.3.4) by first computing $\frac{1}{2}\mathcal{E}_m(fu, v) + \frac{1}{2}\mathcal{E}_m(u, fv) - \frac{1}{2}\mathcal{E}_m(f, uv)$ and then taking the limit as $m \to \infty$.

5.3.2. Show that the Kusuoka measure (5.3.5) is independent of the choice of orthonormal basis.

5.3.3. On the unit disk, the functions $\{r^{|n|}e^{in\theta}\}$ for $n \neq 0$ form an orthogonal basis for the harmonic functions of finite energy modulo constants. Compute the normalizing factor to obtain an orthonormal basis, and then compute the analog of the Kusuoka measure.

5.3.4. Show that (5.3.7) defines a mapping $(u, v) \to f_{u,v}$ from $\operatorname{dom} \mathcal{E} \times \operatorname{dom} \mathcal{E} \to L^1(d\nu)$ that is bilinear and continuous.

5.3.5. Prove the estimate (5.3.13) just using the fact that no nonconstant harmonic function exists with $\mathcal{E}_{F_i K}(h) = \frac{1}{3}\mathcal{E}(h)$ for $i = 0, 1, 2$.

5.3.6.* Find the best constant ε in (5.3.13) by explicitly computing

$$\sum_{i=0}^{2} \left| \frac{1}{3}\mathcal{E}(h) - \mathcal{E}_{F_i K}(h) \right|$$

for all harmonic functions with $\mathcal{E}(h) = 1$.

5.3.7.* Compute $\nu(C)$ on SG for all cells of levels 2 and 3.

5.3.8.* Prove that ν_u is not absolutely continuous with respect to μ for any nonconstant $u \in \operatorname{dom} \mathcal{E}$ on SG.

5.3.9. Show that $\nu \circ F_i^{-1}$ is absolutely continuous with respect to ν, and conversely.

5.4 FRACTAL BLOW-UPS AND FRACTAFOLDS

In this section we discuss two methods to construct global structures built up out of copies of a pcf fractal K. In the first method, the fractal blow-up, we construct a noncompact space which retains the self-similar structure in the large. In the second method, we simply glue together a number of copies of K (finite or infinite) to produce a *fractafold* $\tilde{K}$ with the property that every point $x \in \tilde{K}$ has a neighborhood isomorphic to a neighborhood of K. The fractal blow-up is a special case of a fractafold. With both methods we are able to transfer the structures of energy, measure, and Laplacian.

We start with the self-similar identity

$$(5.4.1) \qquad K = \bigcup_{i=1}^{N} F_i K,$$

choose a value of $i = i_1$, and look at $F_{i_1}^{-1} K$. Note that $K \subseteq F_{i_1}^{-1} K$, and in fact $F_{i_1}^{-1} K$ can be written

$$(5.4.2) \qquad F_{i_1}^{-1} K = \bigcup_{i=1}^{N} F_{i_1}^{-1} \circ F_i K,$$

where the sets $F_{i_1}^{-1} \circ F_k K$ are similar copies of K, which we view as cells of level 0. For the second step in the blow-up we choose i_2 and take $F_{i_1}^{-1} F_{i_2}^{-1} K$, since

$$(5.4.3) \qquad F_{i_1}^{-1} F_{i_2}^{-1} K = \bigcup_{i=1}^{N} F_{i_1}^{-1} F_{i_2}^{-1} F_i K$$

contains $F_{i_1}^{-1} K$ (choose $i = i_2$ on the right). It is clear that

$$(5.4.4) \qquad K \subseteq F_{i_1}^{-1} K \subseteq \cdots \subseteq F_{i_1}^{-1} \cdots F_{i_m}^{-1} K \subseteq \cdots$$

is an increasing sequence of fractals, each similar to K, for any choice of $\{i_1, i_2, \ldots\}$. Thus it makes sense to define the blow-up K_∞ as the union

$$(5.4.5) \qquad K_\infty = \bigcup_{m=1}^{\infty} F_{i_1}^{-1} \cdots F_{i_m}^{-1} K.$$

If C is any m-cell in K we may think of $F_{i_1}^{-1} \cdots F_{i_m}^{-1} C$ as a 0-cell in K_∞. Then K_∞ contains infinitely many 0-cells, and each 0-cell is contained in a unique (-1)-cell, and so on. Thus the structure of K_∞ going toward the infinitely large mimics the structure of K going toward the infinitely small. Of course K_∞ depends on the choice of sequence $\{i_1, i_2, \ldots\}$. If we modify just a finite number of indices, the two blow-ups will be similar, but in general there are an uncountably infinite number of blow-ups that are not even homeomorphic.

We can create a sequence of infinite graphs $\{\Gamma_m\}$, $m \in \mathbb{Z}$, approximating K_∞ as $m \to \infty$. We take Γ_0 to be the graph with vertices V_0 equal to the union of the boundaries of the 0-cells, with two points in V_0 connected by an edge if and only if they belong to the same 0-cell. More generally, V_m consists of the union of the boundaries of all m-cells, and $x \underset{m}{\sim} y$ if and only if x, y belong to the same m-cell. This makes sense for any $m \in \mathbb{Z}$, and the sets V_m are increasing with m. Also,

$$(5.4.6) \qquad V_* = \bigcup_{m=-\infty}^{\infty} V_m$$

is dense in K_∞. In general, an m-cell may be written $F_{i_1}^{-1} \cdots F_{i_n}^{-1} F_{j_1} \cdots F_{j_{m+n}} K$ for some choice of $(j_1, \ldots, j_{m+n})$ with $n \geq -m$. We associate the conductance

$$(5.4.7) \qquad \left(\prod_{\ell=1}^{n} r_{i_\ell} \prod_{k=1}^{m+n} r_{j_k}^{-1} \right) c_{pr}$$

to the edge joining $F_{i_1}^{-1} \cdots F_{i_n}^{-1} F_{j_1} \cdots F_{j_{m+n}} q_p$ and $F_{i_1}^{-1} \cdots F_{i_n}^{-1} F_{j_1} \cdots F_{j_{m+n}} q_r$. Then the energy at level m is defined to be

$$(5.4.8) \qquad \mathcal{E}_m(u) = \sum_{x \underset{m}{\rightarrow} \sim y} c(x, y)(u(x) - u(y))^2 \quad \text{(may be } +\infty\text{)}.$$

It is easy to see that $\mathcal{E}_m(u)$ is increasing in m, so

$$(5.4.9) \qquad \mathcal{E}(u) = \lim_{m \to \infty} \mathcal{E}_m(u)$$

is always well defined, and we define $\text{dom}\,\mathcal{E}$ to be the space of functions with $\mathcal{E}(u) < \infty$. A function is *piecewise harmonic of level m* if u is continuous and the restriction of u to every m-cell is harmonic. In that case $\mathcal{E}_{m'}(u) = \mathcal{E}_m(u)$ for every $m' \geq m$ (this is mainly of interest when $\mathcal{E}_m(u) < \infty$). We can also define harmonic functions (piecewise harmonic of level m for all $m \to -\infty$) on K_∞, but typically there are Liouville-type theorems to the effect that there are no nonconstant harmonic functions of finite energy (or even bounded).

Whether or not K_∞ has a boundary depends on the particular blow-up. If there is a point in $\partial(F_{i_1}^{-1} \cdots F_{i_m}^{-1} K) = F_{i_1}^{-1} \cdots F_{i_m}^{-1}(\partial K)$ for every m, then we consider that point as a boundary point of K_∞; otherwise K_∞ has no boundary. Clearly the only way to get a boundary point is to take all but a finite number of the values i_j equal to one value. Thus, up to similarity, these are the spaces we can represent as $F_i^{-\infty} K$, where $F_i q_i = q_i$ for one of the boundary points of K.

Let μ be any self-similar measure on K with weights $\{\mu_i\}$. Then we can extend μ to an infinite measure μ_∞ on K_∞ by setting

$$(5.4.10) \qquad \mu_\infty(F_{i_1}^{-1} \cdots F_{i_n}^{-1} F_{j_1} \cdots F_{j_{m+n}} K) = \prod_{\ell=1}^{n} \mu_\ell^{-1} \prod_{k=1}^{m+n} \mu_k.$$

We may then define a Laplacian Δ_∞ on K_∞ by the weak formulation $u \in \text{dom}\,\Delta_\infty$ with $\Delta_\infty u = f$ if $u \in \text{dom}\,\mathcal{E}$, f is continuous, and

$$(5.4.11) \qquad \mathcal{E}(u, v) = - \int_{K_\infty} fv\,d\mu_\infty$$

for every $v \in \text{dom}_0\mathcal{E}$ (if K_∞ has no boundary, then $\text{dom}_0\mathcal{E} = \text{dom}\,\mathcal{E}$). In fact, we can give a more general definition requiring only that u belong locally to $\text{dom}\,\mathcal{E}$, but we leave the details to the exercises. Most of the properties of energy and Laplacian on K extend to K_∞ in a routine manner.

The fractal blow-up may be thought of as a collection of copies of K (the 0-cells) that are glued together at boundary points. If we examine a particular junction point x where several 0-cells are joined, then there must exist an n-cell for $n < 0$ which has x as a nonboundary point. So we may conclude that every point in K_∞ has a neighborhood that is similar to a neighborhood of a point in K. It is this property that we use to define a *fractafold* based on K. This just mimics the definition of manifold (or manifold with boundary) where K plays the role of Euclidean space (or Euclidean half-space). The fractal blow-ups are all special cases of fractafolds, but the fractafolds have no requirement that the global structure have any pre-determined form. We can just take a finite or infinite collection of copies of K and glue together some of the boundary points of the copies of K, as long as we model each of these newly created junction points on some junction points already in K. Once we decide on the relative sizes of each of the copies, with respect to both conductances and measure, we can extend the definition of energy and Laplacian from K to the fractafold.

For the rest of this section we consider only the case $K = SG$ with the standard energy and measure, and we assume all the copies of K are of the same size. We are then allowed to pair off any of the boundary points of the copies of K. If all the boundary points are paired, then the fractafold has no boundary; otherwise, the unpaired boundary points of the copies of K are boundary points of the fractafold. The simplest example is the double $\tilde{K}$, discussed in Section 1.1, consisting of two copies of K with corresponding boundary points paired.

We may describe the structure of a fractafold $\mathcal{F}$ by a graph $G_{\mathcal{F}}$ whose vertices are in one-to-one correspondence with the copies of K that make up $\mathcal{F}$, and for each pairing of boundary points of copies of K we create an edge joining the corresponding vertices in $G_{\mathcal{F}}$. Note that we allow more than one edge to join two vertices (this is sometimes called a "multigraph"). It might seem that we are losing information by not saying which boundary ponts are being paired, but in fact the D_3 symmetry of SG means that all three boundary points are isomorphic. Each vertex in $G_{\mathcal{F}}$ has at most three edges, and if $\mathcal{F}$ has no boundary then $G_{\mathcal{F}}$ is a 3-regular graph. Conversely, if G is any 3-regular graph, we can use it as a "blueprint" to construct a fractafold $\mathcal{F}$ without boundary having $G_{\mathcal{F}} = G$.

Suppose that $\mathcal{F}$ is a compact fractafold (only a finite number of copies of K) without boundary. Then $G_{\mathcal{F}}$ is a finite 3-regular graph. It is possible to go from a description of the spectrum of $G_{\mathcal{F}}$ to the spectrum of the Laplacian on $\mathcal{F}$. In particular, if we start with two 3-regular graphs that are isospectral but not isomorphic, then we obtain two fractafolds that are isospectral but not homeomorphic.

If we have a compact fractafold $\mathcal{F}$ with boundary, then we can construct its double $\tilde{\mathcal{F}}$, consisting of two copies of $\mathcal{F}$ glued at corresponding boundary points. Then $\tilde{\mathcal{F}}$ is a fractafold without boundary, so we can determine the spectrum of the Laplacian on $\tilde{\mathcal{F}}$ exactly. It is easy to see that the Dirichlet eigenfunctions on $\mathcal{F}$ extend by odd reflection to eigenfunctions on $\tilde{\mathcal{F}}$, and the Neumann eigenfunctions on $\mathcal{F}$ extend by even reflection to eigenfunctions on $\tilde{\mathcal{F}}$. In this way we can easily pick apart the spectrum of $\tilde{\mathcal{F}}$ to obtain the Dirichlet and Neumann spectra of $\mathcal{F}$. This idea was used implicitly in the discussion of the Neumann spectrum of K in Section 3.3.

The blow-ups K_∞ are noncompact fractafolds, so the Laplacian Δ_∞ does not have a compact resolvant. Nevertheless, at least in the case that K_∞ has no boundary, we can give a complete spectral analysis of Δ_∞ in terms of localized eigenfunctions. Recall that every nonconstant Neumann eigenfunction on K belongs to either the 6-series or the 5-series and has a generation of birth m_0, where it is associated to either a vertex in V_{m_0-1} (6-series) or a loop in Γ_{m_0-1} (5-series). In K_∞ we have similar localized eigenfunctions for all choices of $m_0 \in \mathbb{Z}$. For each choice of $\{\varepsilon_j\}$ and $m_0 = 0$ we obtain an eigenvalue λ, and for other choices of m_0 the eigenvalue becomes $5^{m_0}\lambda$. The eigenspace associated to one such eigenvalue is always infinite dimensional. The localized eigenfunctions defined in Section 3.3 (extended to K_∞) are linearly independent but not orthogonal, but nothing prevents us from using Gram–Schmidt to produce an orthonormal basis of compactly supported eigenfunctions. By combining all these orthonormal sets of functions we obtain a single orthonormal set, since the eigenfunctions associated to different eigenvalues are orthogonal to each other. (A minor point here is that for two distinct choices of λ and λ' for $m_0 = 0$, we never have $\lambda' = 5^m \lambda$ for any $m \in \mathbb{Z}$.) It turns out that this orthonormal set of functions is in fact a basis for $L^2(d\mu_\infty)$.

EXERCISES

5.4.1. Show that a blow-up of SG has a boundary point if and only if all but a finite number of indices i_j are equal. Show that it is contained in a half-plane if and only if one value i occurs only a finite number of times among the indices $\{i_j\}$.

5.4.2. Show that on the blow-up $F_0^{-\infty}(SG)$ with boundary point q_0, the only bounded harmonic functions are the constants.

5.4.3. Formulate a definition of Δ_∞ for functions u that are only locally in $\mathrm{dom}\,\mathcal{E}$ (i.e., the restriction to each subspace $F_{i_1}^{-1} \cdots F_{i_m}^{-1} K$ has finite energy).

5.4.4. Give an equivalent pointwise formula for $\Delta_\infty u$ on V_*.

5.4.5. Show that if $\mathcal{F}$ is a compact fractafold with boundary and $\tilde{\mathcal{F}}$ is its double, then a function is a Dirichlet (resp., Neumann) eigenfuncton of the Laplacian on $\mathcal{F}$ if and only if its odd (resp., even) extension to $\tilde{\mathcal{F}}$ is an eigenfunction of the Laplacian on $\tilde{\mathcal{F}}$.

5.4.6.* Let $\mathcal{F}$ be a compact fractafold without boundary and define the Laplacian on $G = G_\mathcal{F}$ by

$$\Delta u(x) = \sum_{y \sim x}(u(y) - u(x))$$

as usual (G is 3-regular).

(a) Show that 6 is in the spectrum of the Laplacian on G if and only if G is 2-colorable, in which case it has multiplicity 1.

(b) If $\lambda \neq 6$, then λ is in the spectrum of the Laplacian on G if and only if it is in the spectrum of Δ_0 on Γ_0, with equal multiplicities.

(c) Let N be the number of vertices in G. Show that N is even. Show that the multiplicity of 6 in the spectrum of Δ_0 is $\frac{1}{2}N + 1$ is G if 2-colorable, and $\frac{1}{2}N$ if G is not 2-colorable.

(d) Show that 2 is an eigenvalue of Δ_1 on Γ_1 if and only if G is 2-colorable, in which case it has multiplicity 1, but 2 is not an eigenvalue of any Δ_k on Γ_k for $k \geq 2$.

5.4.7. Let $\mathcal{F}$ be a compact fractafold with boundary $\{x_1, \ldots, x_N\}$. Show that the space of harmonic functions on $\mathcal{F}$ is N-dimensional, and each harmonic function is uniquely determined by its boundary values.

5.5 SINGULARITIES

We defined a harmonic function on SG to be continuous, but since the harmonic equation only holds on the interior, it might make sense to allow the function to be discontinuous (or undefined) at the boundary. On the interval this distinction is meaningless, since every linear function on the interior extends continuously to the boundary. But on SG, as we have already seen in Exercise 1.4.7, there exists a function harmonic on $K \setminus \{q_0\}$ for $K = SG$ (continuous on $K \setminus \{q_0\}$ and satisfying $\Delta_m u(x) = 0$ for every $x \in V_m$ except the neighbors of q_0) that has a pole near q_0. This function has infinite energy and is not even integrable. In fact, it is easy to see that the space of harmonic functions on $K \setminus \{q_0\}$ is exactly four-dimensional, spanned by the three-dimensional space of harmonic functions on K and this function with a pole. Every such function is determined by its values at the four points q_1, q_2, F_0q_1, F_0q_2, and these values may be arbitrarily assigned. Loosely speaking, we may refer to these functions as harmonic functions with singularity at q_0, keeping in mind that we allow removable singularities.

What can we say about other kinds of point singularities? We already know an important example: The Green's function $G(\cdot, y)$ is harmonic with singularity at y. This function is in fact continuous, but it is not globally harmonic, so y is not a removable singularity. Of course, this is exactly analogous to the Green's function on the interval, which is a continuous piecewise linear function. But there are other kinds of singlarities as well. In fact, the space of harmonic functions on $K \setminus \{x_0\}$ has dimension 7 if x_0 is a nonboundary junction point and 6 if x_0 is a nonjunction point.

We begin with a simple extension theorem that allows us to localize the study of singularities.

LEMMA 5.5.1 *Let S be any closed set in $F_w K$ that is disjoint from the boundary $F_w V_0$. Then any harmonic function on $F_w K \setminus S$ extends uniquely to a harmonic function on $K \setminus S$.*

Proof: It suffices to prove this for $F_0 K$. The situation is illustated in Figure 5.5.1. Give a harmonic function on $F_0 K \setminus S$, the values c and d are determined. The choice of the values a and b will then determine a harmonic function on $F_1 K \cup F_2 K$, so the issue becomes the matching conditions for normal derivatives at the points F_0q_1 and F_0q_2. This leads to two independent linear equations in a and b, which leads to a unique solution. $\qquad \square$

It is possible to show using the lemma that any harmonic function in $K \setminus \{x_1, \ldots, x_N\}$ may be written as a sum of harmonic functions in $K \setminus \{x_n\}$ for $n = 1, \ldots, N$, and the sum is unique modulo global harmonic functions. The lemma also implies that to understand the harmonic functions on $K \setminus \{x_0\}$ for x_0 a nonboundary junction

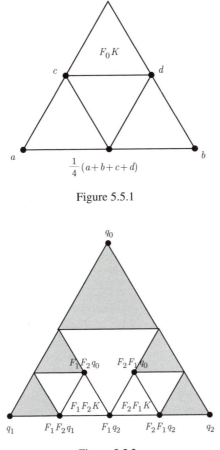

Figure 5.5.1

Figure 5.5.2

point, it suffices to understand the case $x_0 \in V_1$, say $x_0 = F_1 q_2 = F_2 q_1$. In this case we consider the fractafold illustrated by the shaded region in Figure 5.5.2, which may be described as K with the interior of $F_1 F_2 K \cup F_2 F_1 K$ removed, or as the union of the 1-cell $F_0 K$ and four 2-cells. Its boundary consists of seven points (all the black dots in Figure 5.5.2 except $F_1 q_2$).

Any harmonic function on this fractafold is uniquely determined by its values at these points. It can be shown that any harmonic function on this fractafold extends uniquely to a harmonic function on $K \setminus \{F_1 q_2\}$. In Figure 5.5.3 we exhibit four such functions, $\tilde{h}_4, \tilde{h}_5, \tilde{h}_6, \tilde{h}_7$ which, together with the global harmonic functions, span this seven-dimensional space. In addition to the Green's function $\tilde{h}_4$, we have one bounded discontinuous function $\tilde{h}_5$ and two functions with poles, one symmetric ($\tilde{h}_6$) and one skew-symmetric ($\tilde{h}_7$). We leave the details to the exercises.

Next consider the case that x_0 is a nonjunction point. Then there is a unique infinite word $w = (w_1, w_2, \ldots)$ such that x_0 is in the interior of the cell $C_m = F_{w_1} \cdots F_{w_m} K$ for all m. If we take m large enough, then the boundary of C_m is

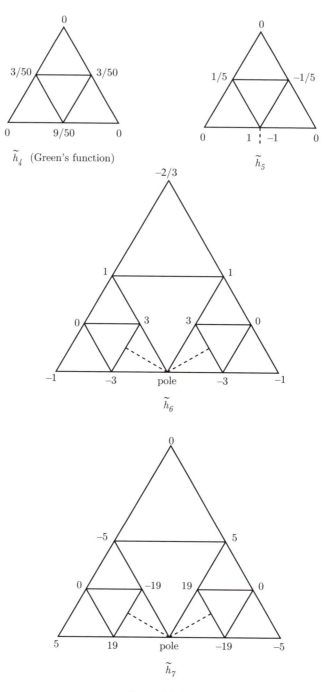

$\widetilde{h}_4$ (Green's function)

$\widetilde{h}_5$

$\widetilde{h}_6$

$\widetilde{h}_7$

Figure 5.5.3

disjoint from V_0. Now consider the fractafold $\mathcal{F}_m$ obtained by deleting the interior of C_m from K. Then $\mathcal{F}_m$ has six boundary points, $V_0 \cup F_{w_1} \cdots F_{w_m} V_0$, and so has a six-dimensional space of harmonic functions. Since $\mathcal{F}_m \subseteq \mathcal{F}_{m+1}$, any harmonic function on $\mathcal{F}_{m+1}$ restricts to a harmonic function on $\mathcal{F}_m$. It is easy to show that the restriction map has zero kernel (see the exercises), and since it is a linear map from a six-dimensional space to a six-dimensional space, it must be onto. In other words, a harmonic function on $\mathcal{F}_m$ extends uniquely to a harmonic function on $\mathcal{F}_{m+1}$. Continuing by induction, every harmonic function on $\mathcal{F}_m$ extends uniquely to a harmonic function on $K \setminus \{x_0\}$. This gives an abstract proof that the space is six-dimensional. A more concrete proof can be given based on the determining data $\{h(q_i), \partial_n h(q_i)\} = \{a_i, n_i\}$, $i = 0, 1, 2$. We can show explicitly how this data determines $\{h(F_{w_1} q_i), \partial_n h(F_{w_1} q_i)\} = \{a_i', n_i'\}$ by solving a set of linear equations. For example, if $w_1 = 0$, then $\{a_0, n_0\} = \{a_0', n_0'\}$. For h to be harmonic at $F_1 q_2$ we must have

$$(5.5.1) \qquad h(F_1 q_2) = \frac{1}{4}\left(a_1 + a_2 + a_1' + a_2'\right).$$

By the definition of normal derivatives,

$$(5.5.2) \qquad \begin{cases} \frac{3}{5} n_1 = 2a_1 - a_1' - \frac{1}{4}\left(a_1 + a_2 + a_1' + a_2'\right), \\ \frac{3}{5} n_2 = 2a_2 - a_2' - \frac{1}{4}\left(a_1 + a_2 + a_1' + a_2'\right). \end{cases}$$

On the other hand, the matching conditions for normal derivatives at $F_0 q_1$ and $F_0 q_2$ yield

$$(5.5.3) \qquad \begin{cases} -\frac{3}{5} n_1' = 2a_1' - a_1 - \frac{1}{4}\left(a_1 + a_2 + a_1' + a_2'\right) \\ -\frac{3}{5} n_2' = 2a_2' - a_2 - \frac{1}{4}\left(a_1 + a_2 + a_1' + a_2'\right). \end{cases}$$

So (5.5.2) and (5.5.3) are necessary and sufficient conditions to have a harmonic function on $\mathcal{F}_1 = F_1 K \cup F_2 K$ with the given boundary values and normal derivatives. But it is easy to see that these equations may be solved uniquely for the primed variables in terms of the unprimed variables (or vice versa). We then repeat the process inductively.

What about singularities of other types of functions? We might expect even more singular poles, but this does not appear to be the case. To be specific, we just look at singularities of polynomials at the boundary point q_0. Not surprisingly, the space $\mathcal{H}^j(K \setminus \{q_0\})$ of solutions of $\Delta^{j+1} u = 0$ on $K \setminus \{q_0\}$ has dmension $4(j + 1)$. In addition to the $3(j + 1)$-dimensional space $\mathcal{H}^j$ of global solutions, spanned by $P_{\ell k}$ for $\ell = 0, 1, \dots, j$ and $k = 1, 2, 3$, we may identify another sequence $P_{\ell 4}$ for $\ell = 0, 1, \dots, j$ of solutions with nonremovable singularities at q_0. P_{04} is just the harmonic function with a pole at q_0 described in Exercise 1.4.7. Just like (5.1.8), we want

$$(5.5.4) \qquad \Delta P_{j4} = P_{(j-1)4},$$

which implies

$$(5.5.5) \qquad P_{j4}(x) = -\int G(x, y) P_{(j-1)4}(y) d\mu(y) + h(x)$$

for some harmonic function h. (Note that when $j = 1$, the integral in (5.5.5) is well defined even though P_{04} has a nonintegrable pole, because G vanishes at $y = q_0$ fast enough to make the product integrable). It turns out that we can choose the harmonic function in (5.5.5) in such a way that we have the scaling identity

$$(5.5.6) \qquad\qquad P_{j4} \circ F_0^n = \left(\frac{3}{5^j}\right)^m P_{j4}$$

(compare with Exercise 5.1.8). The details are too technical to present here. The surprising conclusion is that for $j \geq 1$ there is no pole; in fact, P_{j4} is continuous and vanishes at q_0. We may easily explain this by the "smoothing effect" of integrating against the Green's function. At present, there is no hint of anything like the pole singularities seen in conventional Laurent expansions.

EXERCISES

5.5.1. Show that the condition that u restricted to every cell not containing q_0 is harmonic, is not equivalent to u being harmonic on $K \setminus \{q_0\}$. (Hint: What does the condition say at $F_1 q_2$?)

5.5.2. Show that every harmonic function on $K \setminus \{q_0\}$ is uniquely determined by its values on the points $\{q_1, q_2, F_0 q_1, F_0 q_2\}$. In particular, show that there is a two-dimensional subspace of even functions (resp., odd functions) with respsect to the reflection fixing q_0.

5.5.3. For harmonic functions in Exercise 5.5.2, find a linear relation among the values $\{h(q_1), h(q_2), h(F_0 q_1), h(F_0 q_2)\}$ that is necessary and sufficient for h to extend to a global harmonic function.

5.5.4. Find the explicit solution for the variables a and b in the proof of Lemma 5.5.1.

5.5.5. Verify that the functions shown in Figure 5.5.3 are harmonic on $K \setminus \{F_1 q_2\}$.

5.5.6. Show that a harmonic function on $\mathcal{F}_{m+1}$ whose restriction to $\mathcal{F}_m$ vanishes must vanish on $\mathcal{F}_{m+1}$.

5.5.7. Solve the equations (5.5.2) and (5.5.3) explicitly for the primed variables in terms of the unprimed variables.

5.6 PRODUCTS OF FRACTALS

Euclidean space is constructed by taking products of lines, and differential calculus on Euclidean space may be based on partial derivatives, which are essentially one-dimensional constructions. So we expect to obtain an interesting theory by taking products of fractals and suitably "lifting" the energy and Laplacian from the factors to the product. Indeed, since the theory of differential operators on pcf fractals is more akin to the theory of ordinary differential equations (ODEs), we can hope to obtain some true analogs of PDE theory in the context of products. To keep the discussion simple we will only look at the product of two copies of SG, but similar things can be said for the product of two or more copies of any pcf fractal, or even for a product of different pcf fractals.

We adopt the following notation: K' and K'' are both equal to SG, and $K = K' \times K''$. We will write a variable $x \in K$ as $x = (x', x'')$ for $x' \in K'$ and $x'' \in K''$. In general we will use single primes and double primes to denote any object associated to the first or second factor.

The first simple observation is that K is connected and self-similar but not pcf. In fact, the nine similarities $F_{j'j''} = F_{j'} \otimes F_{j''}$ for j' and j'' ranging over 0, 1, 2 determine the self-similar identity

$$(5.6.1) \qquad\qquad\qquad K = \cup F_{j'j''}K,$$

which is a simple consequence of (1.1.9) on each factor. We think of (5.6.1) as a decomposition of K into 1-cells, so the key question is, How do different 1-cells intersect? If $i' \neq j'$ and $i'' \neq j''$, then $F_{i'i''}K \cap F_{j'j''}K$ is the single point $(F_{i'}q_{j'}, F_{i''}q_{j''})$. On the other hand, if $i' = j'$ but $i'' \neq j''$ then the intersection is $F_{i'}K' \times \{F_{i''}q_{j''}\}$, which is isometric to SG. (Similar story if $i' \neq j'$ but $i'' = j''$.) So this natural decomposition is not finitely ramified. It is not hard to see that no finitely ramified decomposition exists; in other words, K remains connected after the removal of any finite set of points. Also, the natural *boundary* of K consists of the union of the sets $q_{i'} \times K''$ and $K' \times q_{i''}$, and it is clearly not finite. It is hard to visualize K (it naturally embeds in $\mathbb{R}^4$), but if you think of a square, with four quarter-squares that intersect along intervals except for the two diagonal pairs that intersect at the center point, you will have a good analog.

Using the analogy of the square will help us see how to extend the notions of energy and Laplacian to the product. Indeed, the energy on the square is

$$(5.6.2)$$
$$\mathcal{E}(u) = \int_0^1 \int_0^1 |\nabla u(x', x'')|^2 dx' dx''$$
$$= \int_0^1 \left(\int_0^1 \left| \frac{\partial u}{\partial x'}(x', x'') \right|^2 dx' \right) dx'' + \int_0^1 \left(\int_0^1 \left| \frac{\partial u}{\partial x''}(x', x'') \right|^2 dx'' \right) dx'$$
$$= \int_0^1 \mathcal{E}'(u(\cdot, x'')) dx'' + \int_0^1 \mathcal{E}''(u(x', \cdot)) dx',$$

which we may abbreviate

$$(5.6.3) \qquad\qquad\qquad \mathcal{E} = \mathcal{E}' \otimes d\mu'' + d\mu' \otimes \mathcal{E}'',$$

where μ and μ'' are the standard measures on the interval. Note that (5.6.3) reveals that there are actually two ingredients from analysis on the interval, energy and measure, that are used to build the energy on the square. Of course the formula (5.6.2) is not the whole story, since we also need a description of $\text{dom}\,\mathcal{E}$, which necessarily must be a bit complicated, since it is not contained in the continuous functions (so points do not have positive capacity). For example, the function $\log |\log((x' - x_0')^2 + (x'' - x_0'')^2)|$ has finite energy. One acceptable definition is that a measurable function on the square is in $\text{dom}\,\mathcal{E}$ if and only if for almost every x'', $u(\cdot, x'') \in \text{dom}\,\mathcal{E}'$, and for almost every x', $u(x', \cdot) \in \text{dom}\,\mathcal{E}''$, and the integrals

$$\int_0^1 \mathcal{E}'(u(\cdot, x''))dx'' \quad \text{and} \quad \int_0^1 \mathcal{E}''(u(x', \cdot))dx'$$

exist and are finite.

It is easy to mimic this construction on K. First we define a measurable function u to have *minimal regularity* if and only if for almost every x'', $u(\cdot, x'') \in \text{dom}\,\mathcal{E}'$, and for almost every x', $u(x', \cdot) \in \text{dom}\,\mathcal{E}''$. Such a function is in $\text{dom}\,\mathcal{E}$ if and only if

(5.6.4) $$\mathcal{E}(u) = \int_{K''} \mathcal{E}'(u(\cdot, x''))d\mu''(x'') + \int_{K'} \mathcal{E}''(u(x', \cdot))d\mu'(x')$$

exists and is finite. It is then not difficult to show that a function of minimal regularity belongs to $\text{dom}\,\mathcal{E}$ if and only if

(5.6.5) $$\begin{cases} \sup_m \int_{K''} \mathcal{E}_m(u(\cdot, x''))d\mu''(x'') \quad \text{and} \\ \sup_m \int_{K'} \mathcal{E}_m(u(x', \cdot))d\mu'(x') \end{cases}$$

are both finite. For two such functions u and v we have

(5.6.6) $$\mathcal{E}(u, v) = \lim_{m \to \infty} \left(\int_{K''} \mathcal{E}'_m(u(\cdot, x''), v(\cdot, x''))d\mu''(x'') \right.$$
$$\left. + \int_{K'} \mathcal{E}''_m(u(x', \cdot), v(x', \cdot))d\mu'(x') \right).$$

Note, however, that the expressions in (5.6.5) and (5.6.6) are not discrete approximations to energy as they involve integrals. Even in the case of the square, there is no discrete approximation to energy that is valid on all of $\text{dom}\,\mathcal{E}$.

The energy on K satisfies a self-similar identity with renormalization factor $\frac{9}{5}$, namely

(5.6.7) $$\mathcal{E}(u, v) = \sum \left(\frac{9}{5}\right)^{-1} \mathcal{E}\left(u \circ F_{j'j''}, v \circ F_{j'j''}\right).$$

Note that for the square the same identity holds with renormalization factor 1. The fact that these factors are >1 is related to the fact that points have zero capacity. In fact, because $\frac{9}{5} > 1$, functions in $\text{dom}\,\mathcal{E}$ on K may have stronger singularities than for the square: There is a Sobolev embedding theorem $\text{dom}\,\mathcal{E} \subseteq L^p(K)$ (the measure here is $\mu' \times \mu''$) for

(5.6.8) $$p = 2 + \frac{2\log 5}{\log(9/5)},$$

and this result is sharp, as there are functions in $\text{dom}\,\mathcal{E}$ not in $L^{p+\varepsilon}(K)$ for any $\varepsilon > 0$.

One of the simplest classes of functions on K are the *pluriharmonc functions*, defined by the conditions $\Delta'u(\cdot, x'') = 0$ for all x'' and $\Delta''u(x', \cdot) = 0$ for all x'. This is a nine-dimensional space that we may identify with $\mathcal{H}'_l \otimes \mathcal{H}''_l$ with a natural basis

$h_{j'}(x')h_{j''}(x)$ determined by the values on the nine-element *distinguished boundary* $\{(q_{i'}, q_{i''})\}$. We may then build spaces PPH_m of *piecewise pluriharmonic functions of level m*, namely the continuous functions u with $u \circ F_{w'w''}$ piecewise harmonic for all words w' and w'' of length m. Such functions are uniquely determined by their values on $V'_m \times V''_m$. The space $PPH = \cup_m PPH_m$ is dense in dom $\mathcal{E}$. Other basic facts are that a function $u \in$ dom $\mathcal{E}$ has zero energy if and only if it is constant, dom $\mathcal{E}$/constants is a Hilbert space with inner product $\mathcal{E}(u, v)$, dom $\mathcal{E}$ is dense in $L^2(K)$, and the embedding of dom $\mathcal{E}$/constants in $L^2(K)$ is compact. Moreover, $\mathcal{E}$ is local and satisfies the Markov property. Altogether this means that $\mathcal{E}$ is a local, regular Dirichlet form on $L^2(K)$.

Perhaps the best way to understand the energy on K is via double Fourier series. Suppose $\{u'_j\}$ is an orthonormal basis of Neumann eigenfunctions of Δ' on K', with associated eigenvalues $\{\lambda'_j\}$. Then clearly $u_{jk}(x) = u'_j(x')u''_k(x'')$ gives an orthonormal basis for $L^2(K)$. We write

$$(5.6.9) \qquad u = \sum \sum a_{jk}(u)u_{jk}$$

for the expansion of $u \in L^2(K)$ in this basis, with coefficients

$$(5.6.10) \qquad a_{jk}(u) = \int_K u u_{jk} d\mu.$$

Then

$$(5.6.11) \qquad \mathcal{E}(u) = \sum \sum \left(\lambda'_j + \lambda''_k\right) a_{jk}(u)^2$$

and $u \in$ dom $\mathcal{E}$ if and only if the right side of (5.6.11) is finite. Since u'_0 is constant with $\lambda'_0 = 0$, so u_{00} is constant and $a_{00}(u)$ doesn't contribute to (5.6.11).

From the energy and measure $(\mu = \mu' \times \mu'')$ on K, we may define a Laplacian by the weak formulation

$$(5.6.12) \qquad \mathcal{E}(u, v) = -\int_K (\Delta u)v d\mu \quad \text{for all } v \in \text{dom}_0\mathcal{E}.$$

Of course this is formally

$$(5.6.13) \qquad \Delta u = \Delta'u + \Delta''u.$$

It is possible to define dom Δ by requiring that $u \in$ dom $\mathcal{E}$ and Δu be continuous. A more natural space is $\text{dom}_{L^2}\Delta$, where we only require Δu to be in L^2. However, because the boundary of K is so large, it is difficult to give nice characterizations of these domains. The simplest way to get around the problems with the boundary is to consider instead the quadruple cover $\tilde{K} = \tilde{K}' \times \tilde{K}''$, which has no boundary (in the case of the square the quadruple cover is the torus). By tensoring orthonormal bases of eigenfunctions of the Laplacians $\tilde{\Delta}'$ and $\tilde{\Delta}''$ on $\tilde{K}'$ and $\tilde{K}''$ (these bases are obtained by combining Dirichlet and Neumann eigenfunctions on K' extended to be odd and even on $\tilde{K}'$ and renormalized), we obtain an orthonormal basis $\{\tilde{u}_{jk}\}$ of $L^2(\tilde{K})$ of eigenfunctions of $\tilde{\Delta} = \tilde{\Delta}' + \tilde{\Delta}''$ with eigenvalues $\{\tilde{\lambda}_{jk} = \tilde{\lambda}'_j + \tilde{\lambda}'_k\}$. Then it is easy to show that $u \in \text{dom}_{L^2}\tilde{\Delta}$ if and only if

$$(5.6.14) \qquad \sum \sum \left(\tilde{\lambda}'_j + \tilde{\lambda}''_k\right) a_{jk}(u)^2 = \|\tilde{\Delta}u\|^2 < \infty,$$

and

(5.6.15) $$-\tilde{\Delta}u = \sum\sum \left(\tilde{\lambda}'_j + \tilde{\lambda}''_k\right) a_{jk}(u)\tilde{u}_{jk}$$

is the expansion of $-\tilde{\Delta}u$. It can be shown that functions in $\mathrm{dom}_{L^2}\tilde{\Delta}$ are automatically continuous. This is another Sobolev embedding theorem that is valid on products of other pcf fractals, but does not hold for k-fold products with $k > 2$. We may also use (5.6.15) to solve the equation $-\tilde{\Delta}u = f$ for any $f \in L^2(K)$ that is orthogonal to the constants, with the solution unique up to an additive constant. Actually, it is even possible to show that there is a unique solution to $\Delta u = f$ on K with Dirichlet boundary conditions $u|_{\partial K} = 0$ by minimizing

(5.6.16) $$\frac{1}{2}\mathcal{E}(v) + \int_K fv\,d\mu$$

over $v \in \mathrm{dom}_0\mathcal{E}$.

Harmonic functions on K are solutions of $\Delta u = 0$. (There are no nonconstant harmonic functions on $\tilde{K}$, since it has no boundary.) It can be shown that harmonic functions of finite energy minimize energy among functions with the same boundary values, and the harmonic function may be recovered from its boundary values by a kind of Poisson integral formula. Although there is no closed-form formula for the associated Poisson kernel, it can be shown to satisfy the expected estimates.

The expression (5.6.13) for the Laplacian can be generalized to the class of *elliptic* differential operators

(5.6.17) $$a'\Delta'u + a''\Delta''u,$$

where a' and a'' are constants of the same sign. Then a simple modification of (5.6.15) allows us to invert any elliptic operator (modulo constants) on $\tilde{K}$ and stay within $\mathrm{dom}_{L^2}\tilde{\Delta}$. In fact, the inverse operator has many of the properties of pseudodifferential operators. What is more surprising is that there are examples of operators of the form (5.6.17), where a' and a'' have opposite signs, and yet these operators are similarly invertible. We call these operators *quasi-elliptic*. There are no analogs of such operators in classical PDE theory. Indeed, we would expect all operators of the form (5.6.17) with a' and a'' having opposite signs to behave like hyperbolic (or ultrahyperbolic) PDEs, a far cry from the elliptic case.

The explanation for why this is possible is that there are gaps in the set of ratios $\{\tilde{\lambda}'_j/\tilde{\lambda}''_k\}$ of eigenvalues. So if $-a''/a'$ falls into one of these gaps, then there exists $\varepsilon > 0$ such that

(5.6.18) $$\left|a'\tilde{\lambda}'_j + a''\tilde{\lambda}''_k\right| \geq \varepsilon\left(\tilde{\lambda}'_j + \tilde{\lambda}''_k\right).$$

Thus we can solve

(5.6.19) $$-\left(a'\Delta' + a''\Delta''\right)u = f$$

for $f \in L^2(\tilde{K})$ orthogonal to the constants in $\mathrm{dom}_{L^2}\tilde{\Delta}$ by

(5.6.20) $$u = \sum\sum \left(a'\tilde{\lambda}'_j + a''\tilde{\lambda}''_j\right)^{-1} a_{jk}(f)\tilde{u}_{jk}.$$

Using the spectral decimation description of the spectrum of $\tilde{\Delta}'$, it is possible to show that there is a gap containing $\sqrt{5}$ (see the exercises), so $\Delta' - \sqrt{5}\Delta''$ is an example of a quasi-elliptic operator. On the other hand, $\Delta' - \Delta''$ is clearly not an example.

EXERCISES

5.6.1. Show that the function $\log|\log((x'-x_0')^2+(x''-x_0'')^2)|$ has finite energy on the square for any (x_0, x_0'') in the square.

5.6.2. Show that a function of minimal regularity on K is in dom$\mathcal{E}$ (satisfies (5.6.4)) if and only if (5.6.5) are both finite, and show that (5.6.6) holds.

5.6.3. Prove the self-similar identity (5.6.7) and extend it to level m decompositions.

5.6.4. Prove that PPH is dense in the continuous functions in uniform norm and dense in dom$\mathcal{E}$ in the norm $\mathcal{E}(u)^{1/2}+\|u\|_\infty$.

5.6.5. Prove that $\mathcal{E}(u)=0$ if and only if u is constant almost everywhere.

5.6.6. Prove that dom$\mathcal{E}$/constants is a Hilbert space with inner product $\mathcal{E}(u, v)$.

5.6.7. Prove that $\mathcal{E}$ is local and satisfies the Markov property.

5.6.8. Prove that (5.6.11) holds for functions in dom$\mathcal{E}$, and conversely, if the right side of (5.6.11) is finite, then $u \in$ dom$\tilde{\mathcal{E}}$.

5.6.9. Prove that (5.6.15) holds for $u \in \text{dom}_{L^2}\tilde{\Delta}$, and moreover, this domain is characterized by (5.6.14).

5.6.10. Prove that a minimizer of (5.6.16) over $\text{dom}_0\mathcal{E}$ solves $\Delta u = f$ on K.

5.6.11. Show that if (5.6.18) holds then (5.6.20) belongs to $\text{dom}_{L^2}\tilde{\Delta}$ and solves (5.6.19) for any f orthogonal to the constants.

5.6.12.* Let λ' and λ'' be two Neumann eigenvalues, and let $\{\lambda_m'\}$ and $\{\lambda_m''\}$ denote the discrete eigenvalues associated to them by the spectral decimation method.

(a) Show that
$$\frac{\lambda'}{\lambda''} = \lim_{m\to\infty} \frac{\lambda_m'}{\lambda_m''}.$$

(b) Show that there exists m_1 such that $\lambda_{m_1}' \in [3, 5]$ and $\lambda_{m+1}' = \varphi_-(\lambda_m')$ for every $m \geq m_1$, where $\varphi_\pm(x) = \frac{5\pm\sqrt{25-4x}}{2}$, and similarly there exists m_2 for λ''.

(c) If $m_1 = m_2$ show that
$$\frac{\lambda_m'}{\lambda_m''} \leq \frac{\varphi_-^n(5)}{\varphi_-^n(3)}$$
for $n = m - m_1$, while if $m_2 > m_1$ then
$$\frac{\lambda_m'}{\lambda_m''} \geq \frac{\varphi_-^n(3)}{\varphi_-^{n+1}(5)}.$$

(d) Conclude that λ'/λ'' never lies in the interval $[a, b]$ for
$$a = \lim_{n\to\infty} \frac{\varphi_-^n(5)}{\varphi_-^n(3)} \approx 2.0611106$$

and

$$b = \lim_{n\to\infty} \frac{\varphi_-^n(3)}{\varphi_-^{n+1}(5)} \approx 2.428766.$$

Show that $ab = 5$ so $\sqrt{5}$ lies in this gap.

5.7 SOLVABILITY OF DIFFERENTIAL EQUATIONS

Let K be a pcf fractal with Laplacian Δ. We can then build more general differential operators out of Δ. For example, if $u(x) = (u_1(x), \ldots, u_n(x))$ is a function from K to $\mathbb{R}^n$ and $F(x, u)$ is a continuous function from $K \times \mathbb{R}^n$ to $\mathbb{R}^n$, we may consider the system

$$(5.7.1) \qquad\qquad -\Delta u(x) = F(x, u(x))$$

of nonlinear differential equations. Just as in the case of ODEs, we may reduce many higher order equations to the form (5.7.1) by introducing more variables. In analogy with ODEs, we might expect local existence and uniqueness under a Lipschitz condition on F, and even local existence (without uniqueness) in general, without any smoothness assumptions on F, and global existence if F is linear in u. It turns out that we can realize the first two expectations, but not the third.

First we need to state the precise form of the Lipschitz condition, which is actually a local Lipschitz condition in the u-variable alone. Since u takes values in $\mathbb{R}^n$, this is identical to the statement for ODEs:

$$(5.7.2) \qquad \begin{cases} \text{for every } T > 0 \text{ there exists } M_T < \infty \text{ such that} \\ |F(x, u) - F(x, u')| \leq M_T |u - u'| \text{ provided } |u|, |u'| \leq T. \end{cases}$$

For local solvability we will ask that (5.7.1) hold on an m-cell $F_w K$ with $|w| = m$ and m large enough. Unlike the ODE case, we will specify boundary conditions

$$(5.7.3) \qquad\qquad u(F_w q_i) = a_i, \qquad i = 1, \ldots, N_0,$$

on the boundary $F_w V_0$ of $F_w K$, rather than initial conditions.

THEOREM 5.7.1 *Given F satisfying (5.7.2), for every A there exists m such that for all choices of $\{a_i\}$ with $|a_i| \leq A$, the equation (5.7.1) on $F_w K$ (for any $|w| = m$) with boundary conditions (5.7.3) has a unique solution.*

Proof: We give the proof for all $a_i = 0$, leaving the general case to the exercises. If we write $v = u \circ F_w$, then (5.7.1) becomes (after the change of variable $x \to F_w x$)

$$(5.7.4) \qquad\qquad -\Delta v(x) = r_w \mu_w F(F_w x, v(x)) \quad \text{on } K$$

and (5.7.3) says v vanishes on V_0. It follows that (5.7.4) and the boundary condition are equivalent to the integral equation

$$(5.7.5) \qquad\qquad v(x) = r_w \mu_w \int_K G(x, y) F(F_w y, v(y)) d\mu(y),$$

where $G(x, y)$ is the Green's function. If we define $\mathcal{G}v(x)$ to be the right side of (5.7.5), then we are just looking for a fixed point of $\mathcal{G}$, so it suffices to show that $\mathcal{G}$ satisfies the hypotheses of the contractive mapping principle on a suitable ball $\mathcal{B} = \{\|v\|_\infty \leq T\}$ in the Banach space of continuous functions v. Let G_0 denote an upper bound for $G(x, y)$ on $K \times K$ (it is continuous and bounded), and let F_T be an upper bound for $|F(x, u)|$ for $x \in K$ and $|u| \leq T$. Then

$$|\mathcal{G}v(x)| \leq r_w \mu_w G_0 F_T \quad \text{for } v \in \mathcal{B},$$

so by taking m large enough that $r_w \mu_w G_0 F_T \leq T$ we conclude that $\mathcal{G}$ maps $\mathcal{B}$ to itself. Then for v and v' in $\mathcal{B}$ we have

$$|\mathcal{G}v(x) - \mathcal{G}v'(x)| \leq r_w \mu_w G_0 M_T \|v - v'\|_\infty$$

by (5.7.2), and this shows that $\mathcal{G}$ is contractive on $\mathcal{B}$ provided that we take m large enough so that $r_w \mu_w G_0 M_T < 1$.   □

The analog of the Peano existence theorem, local existence to (5.7.1) and (5.7.3) without the assumption (5.7.2), is also valid, but we will not discuss the proof since it is more technical, and the result is of less interest because it is nonconstructive.

Next we consider the question of global solvability for linear equations for $K = SG$. The simplest case is the single equation

(5.7.6) $-\Delta u = \lambda u + f,$

where λ is a constant and f is a continuous function. If λ is not a joint D–N eigenvalue, then it is easy to solve (5.7.6) explicitly by expanding in an orthonormal basis of either Dirichlet or Neumann eigenfunctions. But in the case that λ is a joint D–N eigenvalue it turns out that there are choices of f for which (5.7.6) has no global solutions. In fact, this will be the case for f equal to a λ-eigenfunction. This means that u must be a generalized eigenfunction,

(5.7.7) $(-\Delta - \lambda)^2 u = 0.$

Let n denote the multiplicity of the space of λ-eigenfunctions. For the joint D–N eigenvalues this is typically greater than 3. If we could always solve (5.7.6) then the space of solutions of (5.7.7) would have dimension at least $2n$, and within this space we could impose either Dirichlet or Neumann boundary conditions to obtain a space of dimension $2n - 3$. But the spectral theorem implies that solutions of the generalized eigenfunction equation (5.7.7) satisfying either Dirichlet or Neumann boundary conditions are in fact λ-eigenfunctions. This means $2n - 3 \leq n$, or $n \leq 3$. In fact, it is possible to exhibit explicit eigenfunctions f for which (5.7.6) has no global solutions, but the details are quite technical.

Another interesting problem that is more reminiscent of PDE theory is the solvability of

(5.7.8) $-\Delta u = f \quad \text{on } \Omega,$

for Ω an open subset of K not containing any points of V_0, where f is any continuous function on Ω. The point is that we do not want to make any assumptions on the behavior of f near the boundary of Ω, so in particular f need not extend to a continuous function on K. Results like this in $\mathbb{R}^n$ (for elliptic constant coefficient operators) are due to Malgrange and Ehrenpreis. It turns out that we can prove the solvability of (5.7.8) for any Laplacian on a pcf fractal K.

We give a rough outline of the proof. We write Ω as an infinite union of cells

(5.7.9) $\Omega = \bigcup_{w \in W} F_w K$

that only overlap at boundary points, such that each cell is maximal (not contained in any larger cell in Ω). We may assume that Ω is connected, since it suffices to solve (5.7.8) on each connected component. On each cell $F_w K$ in (5.7.9) we can solve (5.7.8) with Dirichlet boundary conditions and continuously glue together these local solutions to obtain a function v. Note that v is not a solution to (5.7.8) because it doesn't satisfy the matching conditions for normal derivatives at the boundary points of the cells. Call this set of points $\tilde{V}$. Let h be any piecewise harmonic spline with respect to the decompostion (5.7.9) (we are stretching the definition because the decomposition is infinite). Clearly h is uniquely determined by its values on $\tilde{V}$. Then $u = v + h$ will satisfy (5.7.8) at every point of $\Omega \setminus \tilde{V}$, so it suffices to show that h may be chosen so that u satisfies the matching conditions at all the points in $\tilde{V}$.

Now we construct a graph $\tilde{\Gamma}$ with vertices $\tilde{V}$, with edges $x \sim y$ if and only if x and y belong to $F_w V_0$ for some $w \in W$, and assign conductance $c(x, y)$ from Γ_m for $m = |w|$ to this edge. The sum of the normal derivatives of h at x may be written

$$
(5.7.10) \qquad\qquad \sum_{y \sim x} c(x, y)(h(x) - h(y)),
$$

which we may regard as a graph Laplacian $-\tilde{\Delta}\tilde{h}(x)$, where $\tilde{h}$ denotes the restriction of h to $\tilde{V}$. Thus the matching conditions for normal derivatives of u take the form

$$
(5.7.11) \qquad\qquad\qquad\qquad -\tilde{\Delta}\tilde{h} = \tilde{g}
$$

for some prescribed function $\tilde{g}$ on $\tilde{\Gamma}$.

To solve (5.7.11) we approximate the infinite graph $\tilde{\Gamma}$ by a sequence $\{\tilde{\Gamma}_n\}$ of finite connected graphs with boundary ($\partial\tilde{\Gamma}_n$ consists of vertices connected in $\tilde{\Gamma}$ to a vertex not in $\tilde{V}_n$). By prescribing values on $\partial\tilde{\Gamma}_n$ we may solve (5.7.11) on the interior of $\tilde{\Gamma}_n$, and any solution on $\tilde{\Gamma}_{n'}$ restricts to a solution on $\tilde{\Gamma}_n$ for $n < n'$. Although not every solution on $\tilde{\Gamma}_n$ extends to $\tilde{\Gamma}_{n'}$, if we let $E_{n,n'}$ denote the affine subspace of solutions that do extend, then we obtain a nested sequence (for n fixed and n' varying). Such a sequence must stabilize, so $E_{n,n'} = E_n$ for all n' large enough. Starting in E_n we may extend the solution to all of $\tilde{\Gamma}$.

EXERCISES

5.7.1. Complete the proof of Theorem 5.7.1 to the case of general $\{a_i\}$. (Hint: Let $w = u - h$, where h is the harmonic function with the same boundary values as u, and find the differential equation satisfied by w.)

5.7.2. Solve (5.7.6) when λ is not a joint D–N eigenvalue, and show that the same solution is valid if λ is a joint D–N eigenvalue but f is orthogonal to the joint D–N eigenspace.

5.7.3. Show that the decomposition (5.7.9) holds where $\{F_w K\}$ are chosen to be the maximal cells in Ω.

5.8 HEAT KERNEL ESTIMATES

If Δ is some Laplacian in the space variable x, then the heat equation is the space–
time equation

$$(5.8.1) \qquad\qquad \frac{\partial u(x,t)}{\partial t} = c\Delta u(x,t) \quad \text{for } t > 0,$$

where c is a positive constant that we will take to be $c = 1$ for convenience (in the
probability literature, the convention is often $c = \frac{1}{2}$). The initial condition

$$(5.8.2) \qquad\qquad u(x,0) = f(x)$$

in a suitable sense (u does not have to be continuous at $t = 0$ in general) together
with the heat equation means that formally

$$(5.8.3) \qquad\qquad u(x,t) = e^{t\Delta} f(x).$$

If the Laplacian comes with boundary conditions to make it a negative self-adjoint
operator, then those same boundary conditions are imposed on u. If the Laplacian
has a discrete spectrum, with a complete orthonormal basis $\{u_j\}$ of eigenfunctions

$$(5.8.4) \qquad\qquad -\Delta u_j = \lambda_j u_j,$$

then (5.8.3) means

$$(5.8.5) \qquad u(x,t) = \sum_j e^{-t\lambda_j} \left(\int f(y)u_j(y)d\mu(y) \right) u_j(x),$$

and usually the sum and integral can be interchanged to yield

$$(5.8.6) \qquad\qquad u(x,t) = \int p(t,x,y) f(y)d\mu(y),$$

where

$$(5.8.7) \qquad\qquad p(t,x,y) = \sum_j e^{-t\lambda_j} u_j(x)u_j(y)$$

is called the *heat kernel*. (Even if the Laplacian does not have a discrete spectrum, it
is possible to define a heat kernel via the spectral theorem generalization of (5.8.7)
so that the representation (5.8.6) is still valid.) Typically the heat kernel is positive
valued and smooth.

The study of the heat equation, the heat kernel, and heat kernel estimates has
long been an important part of analysis for a wide variety of Laplacians. In recent
decades it has taken on added significance because of many works showing that,
starting with heat kernel estimates, it is possible to derive many desirable consequences. We might paraphrase this philosophy as: "Whatever you want, first get
heat kernel estimates."

In this section we consider the case where the Laplacian is either a Dirichlet or
Neumann Laplacian on a pcf fractal K, or a Laplacian on the double cover $\tilde{K}$ (no
boundary). We will write $p_D(t,x,y)$, $p_N(t,x,y)$, and $\tilde{p}(t,x,y)$ when we need to

distinguish these cases, and $p(t, x, y)$ when we want to make statements about all of them. First we will discuss the standard Laplacian on $K = SG$.

To begin, we note that it is relatively easy to say something about the trace of the heat kernel

$$(5.8.8) \qquad h(t) = \int p(t, x, x) d\mu(x),$$

since by (5.8.7) this becomes

$$(5.8.9) \qquad h(t) = \sum_j e^{-t\lambda_j}$$

and so depends only on the eigenvalues, not the eigenfunctions. We can then relate $h(t)$ to the eigenvalue counting function $\rho(x)$ defined by (3.5.1) via

$$(5.8.10) \qquad h(t) = \int_0^\infty t e^{-ts} \rho(s) ds.$$

Since we know the asymptotic behavior of $\rho(x)$ as $x \to \infty$, we may deduce the behavior of $h(t)$ as $t \to 0^+$. (Note that we are not interested in the behavior of the heat kernel as $t \to \infty$, since it decays exponentially in the Dirichlet case, or goes to a constant in the other cases.) We can rewrite (3.5.5) as

$$(5.8.11) \qquad \rho(x) = g(\log x) x^\beta + o(x^\beta) \quad \text{as } x \to \infty$$

for $\beta = \log 3 / \log 5$, where g is bounded, bounded away from zero, and periodic of period $\log 5$, but not continuous. It follows that

$$(5.8.12) \qquad h(t) = t^{-\beta} \gamma(t) + o(t^{-\beta}) \quad \text{as } t \to 0^+,$$

where $\gamma(t)$ is a continuous function that is bounded, bounded away from zero, multiplicatively periodic of period 5 ($\gamma(5t) = \gamma(t)$), and not constant. In fact,

$$(5.8.13) \qquad \gamma(t) = \int_0^\infty e^{-s} s^\beta g(\log s - \log t) ds.$$

If we expand $g(s)$ in a Fourier series

$$(5.8.14) \qquad g(s) = \sum_{-\infty}^\infty c_n e^{2\pi i n s / \log 5},$$

then we have the Fourier series representation of $\gamma(e^x)$ given by

$$(5.8.15) \qquad \gamma(e^x) = \sum_{-\infty}^\infty c_n \Gamma\left(1 + \beta + \frac{2\pi i n}{\log 5}\right) e^{-2\pi i n x / \log 5}.$$

All these results follow by routine analysis by plugging (5.8.11) into (5.8.10). We leave the verification to the exercises. The significance of (5.8.15) is that the Γ factors decrease rapidly as $n \to \pm\infty$, so γ is smooth even though g is only bounded. On the other hand, since g is not constant, at least some c_n coefficients are nonzero for $n \neq 0$, and that shows that γ is not constant. Also, the terms in (5.8.15) with $n = \pm 1$ can be expected to be considerably larger than all subsequent terms (since we have no real information about the c_n coefficients, this cannot be made into a precise statement), so we expect the graph of $\gamma(e^x)$ to resemble a sine curve $(a + b \sin \frac{2\pi}{\log 5}(x - c)$ for some $a, b, c)$. Numerical calculations confirm this observation.

As mentioned at the end of Section 3.5, the exponent $\beta = \log 3/\log 5$ should be interpreted as the ratio $d/(d + 1)$, where $d = \log 3/\log(5/3)$ is the Hausdorff dimension of SG in the resistance metric, and $d + 1 = \log 5/\log(5/3)$ is the order of the Laplacian. This is consistent with the dimension/order exponent in standard heat kernel estimates.

On-diagonal estimates of the heat kernel show the same power growth in t. We have upper bounds

$$(5.8.16) \qquad\qquad p(t, x, x) \le c_1 t^{-\beta}, \qquad 0 < t \le 1,$$

in all three cases, and lower bounds

$$(5.8.17) \qquad\qquad p(t, x, x) \ge c_2 t^{-\beta}, \qquad 0 < t \le 1,$$

for the Neumann or double cover case. (Obviously (5.8.17) cannot hold in the Dirichlet case since $p_D(t, x, x)$ vanishes for x on the boundary.) There is some experimental evidence that $p(t, x, x)t^\beta$ has the same oscillatory behavior as $\gamma(t)$, at least when x is a junction point.

Off-diagonal estimates $(x \ne y)$ are expected to show a sharp drop-off as x and y separate. In fact, the precise behavior is

$$(5.8.18) \qquad\qquad t^{-\beta} \exp\left(-c \left(\frac{R(x, y)^{d+1}}{t}\right)^{\frac{1}{d_w - 1}}\right),$$

where $d_w = \frac{\log 5}{\log 2}$ is interpreted as a "walk dimension in a shortest path metric." Once again there is an upper bound of $p(t, x, y)$ by a multiple of (5.8.18) in all cases, and a similar lower bound in the Neumann and double cover cases. However, the constant in (5.8.18) is not the same in the upper and lower bounds. The proofs of these estimates use probability theory. Similar estimates are known to hold for nested fractals and some generalizations of nested fractals, with the appropriate values of d and d_w.

One important idea we can easily extract from (5.8.18) is the size of distance $R(x, y)$ relative to t that is required to see the spatial drop-off in the heat kernel. If

$$(5.8.19) \qquad\qquad R(x, y) \le t^{1/(d+1)},$$

then the exponential part of (5.8.18) is bounded below. So distances of this order of magnitude do not result in any spatial decay. But once the distance exceeds $t^{1/(d+1)}$, we see an exponential decay as the distance increases. Since $d > 1$, this cut-off point is much larger than $t^{1/2}$, the value for the usual Gaussian heat kernel in Euclidean space. For this reason the estimate (5.8.18) is called *sub-Gaussian*. However, for many important applications of heat kernel estimates, these sub-Gaussian estimates are quite adequate, and the exact exponents in the exponential part of (5.8.18) are not of great significance.

One very nice property of heat kernels is that they behave very simply when we consider products: The heat kernel on a product is the product of the heat kernels on the factors. Reverting to the notation of Section 5.6, if $K = K' \times K''$ and $p'(t, x', y')$ and $p(t, x'', y'')$ denote heat kernels on K' and K'' while $p(t, x, y)$ denotes the heat kernel (of the same type) on K, then

$$(5.8.20) \qquad\qquad p(t, x, y) = p(t, x', y')p(t, x'', y'').$$

This is a simple consequence of (5.8.7) on each factor and the tensor product construction of eigenfunctions on K. The analog of (5.8.18) sees the value of β double, interpreted simply as doubling the dimension while keeping the order of the operator the same, while the exponential term involves the distance

$$(5.8.21) \qquad d(x, y) = \left(R\left(x', y'\right)^{\frac{d+1}{d_w-1}} + R\left(x'', y''\right)^{\frac{d+1}{d_w-1}} \right)^{\frac{d_w-1}{d+1}}.$$

We note that there is no resistance metric on K since points have zero capacity.

The heat kernel may be used to construct other functions of the Laplacian, for example, the Green's function

$$(5.8.22) \qquad\qquad G(x, y) = \int_0^\infty p_D(t, x, y)dt.$$

For the case of SG this is not very interesting because we already have an explicit construction of $G(x, y)$, but in the product setting it gives quite a nice result. In that case, integrating against $G(x, y)$ gives the unique solution of $-\Delta u = f$ with $u|_{\partial K} = 0$. By substituting the (5.8.18)-type estimate in (5.8.22) for $0 < t \leq 1$ and using crude estimates for $t > 1$, we can show

$$(5.8.23) \qquad\qquad |G(x, y)| \leq cd(x, y)^{1-d}.$$

Thus the Green's function now has a singularity on the diagonal of pseudo-differential operator type.

EXERCISES

5.8.1. Prove (5.8.10).

5.8.2. Show that (5.8.11) implies (5.8.12), with γ given by (5.8.13).

5.8.3. Show the relationship (5.8.15) between the Fourier series of $\gamma(e^x)$ and the Fourier series (5.8.14) of $g(s)$. Then use estimates for the gamma function to show that γ is C^∞.

5.8.4. Prove the product formula (5.8.20).

5.8.5. Prove (5.8.22).

5.8.6.* Prove (5.8.23).

5.8.7.* Use the heat kernel estimates on the quadruple cover $\tilde{K} = \tilde{K}' \times \tilde{K}''$ to prove the Sobolev embedding theorems dom $\tilde{\mathcal{E}} \subseteq L^q(\tilde{K})$ for $q = \frac{4d}{d-1}$, and dom$_{L^2}\tilde{\Delta} \subseteq C(\tilde{K})$.

5.9 CONVERGENCE OF FOURIER SERIES

Convergence of ordinary Fourier series comes with a price: You have to either assume some sort of smoothness for the function, use a summability method, or weaken the notion of convergence. On SG it is virtually free; the only requirement is that you choose the partial sums in a sensible way.

Consider, for example, the Fourier cosine expansion of a continuous function on I:

$$(5.9.1) \qquad f(x) = \sum_{n=0}^{\infty} a_n(f) \cos n\pi x,$$

for the appropriate coefficients. The partial sums

$$(5.9.2) \qquad s_N f(x) = \sum_{n=0}^{N} a_n(f) \cos n\pi x$$

are given by integrating f against a Dirichlet kernel which does not behave like a nice approximate identity. In particular, there are continuous functions for which $s_N f$ does not converge uniformly to f, and there are even points x where $s_N f(x)$ is unbounded. All these troubles disappear if f has some mild smoothness. Also, $s_N f$ converges to f in L^2 norm, and by Carleson's theorem, $s_N f(x)$ converges to f for almost every x.

Another way to obtain convergence is to introduce a summability factor $A_{N,n}$ (with $A_{N,n} \to 1$ as $N \to \infty$ for each n) and define

$$(5.9.3) \qquad \sigma_N f(x) = \sum_{n=0}^{N} A_{N,n} a_n(f) \cos n\pi x.$$

Under suitable conditions on the summability factors, $\sigma_N f \to f$ uniformly as $N \to \infty$ for all continuous functions. The most common choice is Fejér summability, which leads to a nonnegative Fejér kernel, but this is not the best choice. A better choice is to take $A_{N,n} = 1$ for $n < \frac{N}{2}$ and then decrease linearly to zero at $n = N+1$. For this choice, $\sigma_N f = f$ for N large enough if f is a trigonometric polynomial ((5.9.1) is a finite sum).

These well-known results about convergence and nonconvergence are somewhat technical. However, there is one result that is immediately apparent: The Dirichlet kernel cannot satisfy the strong approximate identity estimate

$$(5.9.4) \qquad \int_{|y-x| \geq \varepsilon} |D_N(x, y)| dy \to 0 \quad \text{as } N \to \infty$$

for fixed $\varepsilon > 0$. Indeed, if the integral in (5.9.4) were small for some N, then adding the next term $2\cos(N+1)\pi x \cos(N+1)\pi y$ would make it not small for $N+1$. (Of course, this in itself does not imply the failure of uniform convergence; you need something like the unboundedness of the Lebesgue constants $\int |D_N(x, y)| dy$.) The interesting thing is that this argument, which is really quite generic, does not preclude the possibility that something like (5.9.4) holds along some sequence of values $N_k \to \infty$. For ordinary Fourier series, there is no obvious sequence of values N_k that would make a natural choice.

But now look at the situation for SG. As before, let $\{u_j\}$ denote an orthonormal basis of Neumann eigenfunctions with eigenvalues $\{\lambda_j\}$ increasing. It is very natural to take a partial sum of the Fourier series expansion up to the first $N_m = \#V_m = \frac{3^{m+1}+3}{2}$ eigenvalues, for these correspond to the eigenvalues of Δ_m extended by spectral decimation in the smallest way possible. The highest eigenvalue corresponds to the extension of 6 as the eigenvalue of Δ_m. This must extend to 3 as the eigenvalue of Δ_{m+1}, and then to the eigenvalue

$$(5.9.5) \qquad \lambda_{N_m} = \frac{3}{2} \cdot 5^{m+1} \lim_{n \to \infty} 5^n \varphi_-^{(n)}(3)$$

of Δ. What is the next highest eigenvalue of Δ? It comes from the eigenvalue 5 of Δ_m extended to $\varphi_+(5)$ as the eigenvalue of Δ_{m+1}, and then to the eigenvalue

$$(5.9.6) \qquad \lambda_{N_m+1} = \frac{3}{2} 5^{m+1} \lim_{n \to \infty} 5^n \varphi_-^{(n)}(\varphi_+(5))$$

of Δ. It is clear by inspection that the ratio $\frac{\lambda_{N_m+1}}{\lambda_{N_m}}$ is independent of m. This means that there is a significant gap between λ_{N_m} and λ_{N_m+1}, meaning that the difference is a fixed multiple of λ_{N_m}. Nothing like this happens for ordinary Fourier series, where $\lambda_n = (\pi n)^2$, so $\lambda_{n+1} - \lambda_n = \pi^2(2n+1)$ is small compared with λ_n.

Now the claim is that the partial sums $s_{N_m} f$ are identical to certain approximations $\sigma_{N_m} f$ obtained from summability factors

$$(5.9.7) \qquad A_{N_m,n} = \psi\left(\frac{\lambda_n}{\lambda_{N_m}}\right),$$

where $\psi(t) = 1$ for $0 \leq t \leq 1$ and $\psi(t)$ decreases to zero by $t = \frac{\lambda_{N_m+1}}{\lambda_{N_m}}$, but ψ may be taken to be smooth enough (even C^∞). By analogy with ordinary Fourier series we would expect $\sigma_{N_m} f$ to converge uniformly to f as $m \to \infty$, for any continuous function f. If this is the case, then we would immediately obtain that $s_{N_m} f$ converges uniformly to f. In other words, partial sums converge uniformly as long as we sum "up to a gap."

In fact, the uniform convergence of $\sigma_{N_m} f$ to f is a consequence of a generic transplantation theorem of Duong, Ouhabaz, and Sikora that says that estimates that are true in Euclidean space remain true for functions of the Laplacian, provided the associated heat kernel satisfies sub-Gaussian estimates. And these estimates do hold, as we have seen in Section 5.8.

We may summarize the above discussion as follows.

THEOREM 5.9.1 *Let $\{N_m\}$ be any sequence of integers such that $(\lambda_{N_m+1}/\lambda_{N_m}) - 1$ is bounded away from zero. Then $s_{N_m} \to f$ uniformly as $m \to \infty$ for any continuous function f.*

There are many gaps in the spectrum of the Laplacian on SG. In a certain sense it is asymptotic at infinity to the Julia set of $x(5-x)$ in a neighborhood of zero, which has the structure of a Cantor set. The numerical value of the ratio $\lambda_{N_m+1}/\lambda_{N_m}$ for the values given in (5.9.6) and (5.9.5) is about 1.271. But this is not the largest gap. If we go to the distinct eigenvalue just below (5.9.5) it comes from extending the eigenvalue 5 of Δ_m in the minimal way, namely

$$(5.9.8) \qquad \frac{3}{2} 5^m \lim_{n \to \infty} 5^n \varphi_-^{(n)}(5).$$

This yields a numerical value of about 2.425 for the ratio. The fact that this value is greater than 2 is significant.

Consider the product of two copies of SG, $K = K' \times K''$, as discussed in Section 5.6. Then there are two ways to take partial sums of the eigenfunction expansion (5.6.9): We can take "square" partial sums

$$(5.9.9) \qquad s_N f = \sum_{j=1}^{N} \sum_{k=1}^{N} a_{jk}(f) u_{jk},$$

or "circular" partial sums

$$(5.9.10) \qquad \tilde{s}_M f = \sum_{\lambda'_j + \lambda''_k \leq M} a_{jk}(f) u_{jk}.$$

From the point of view of the spectrum of Δ, the circular partial sums are natural, while the square partial sums are artificial. However, it is easy to see that the square partial sums enjoy the same uniform convergence for any sequence N_m satisfying the hypotheses of Theorem 5.9.1. On the other hand, if we choose N_m so that the ratio is greater than 2, then clearly

$$(5.9.11) \qquad s_{N_m} f = \tilde{s}_{2\lambda_{N_m}} f,$$

so we have uniform convergence along a sequence of circular partial sums. Note, however, that the same reasoning will not work on products of more than two copies of SG because there are no gaps with ratio exceeding 3.

Is it possible to obtain convergence results like the above on other pcf fractals? To do so we need two ingredients: heat kernel estimates and spectral gaps. As mentioned in Section 5.8, heat kernel estimates are known for nested fractals. There is numerical evidence for spectral gaps for the pentagasket, but as yet no proof. One class of fractals where both heat kernel estimates and the existence of spectral gaps are known is the class of homogeneous hierarchical gaskets discussed in Section 4.6.

On the negative side, it is known that spectral gaps are only possible in the "lattice case." Consider a Laplacian Δ_μ as defined in Section 4.4 with respect to a self-similar measure. Then the analog of the scaling identity (2.1.7) is

$$(5.9.12) \qquad \Delta_\mu(u \circ F_j) = r_j \mu_j (\Delta_\mu u) \circ F_j.$$

If all the scaling factors $r_j \mu_j$ are integer powers of a single number, then we say we are in the *lattice case*. (Equivalently, the numbers $\log(r_j \mu_j)$ lie on a lattice.) Otherwise, we are in the *nonlattice case*. A general result of Kigami and Lapidus gives the asymptotics of the eigenvalue counting function analogous to (3.5.5) with the appropriate power of x. In the lattice case the function $g(t)$ is periodic and discontinuous, but in the nonlattice case it is constant. (The nonlattice case is thus completely analogous to the Weyl asymptotic law in smooth analysis, with a different power.) However, it is clear that spectral gaps force $g(t)$ to be nonconstant, hence they can only occur in the lattice case.

EXERCISES

5.9.1. Prove the analog of Theorem 5.9.1 in the product setting for the square partial sums (5.9.9).

5.9.2. Prove (5.9.11) when the ratio $\lambda_{N_m+1}/\lambda_{N_m}$ exceeds 2.

5.9.3. Prove the scaling identity (5.9.12).

5.10 NOTES AND REFERENCES

The definition of spline spaces and the different bases for $\mathcal{H}_k$ in Section 5.1 is from [Strichartz & Usher 2000]. The rest of the material in Section 5.1 is from [Needleman et al. 2004]. Local symmetries were first discussed with respect to the pentagasket in [Adams et al. 2003], but the fact that certain eigenfunctions on SG are constant along line segments was shown directly in [Dalrymple et al. 1999]. The basic properties of energy measures, and the fact that they are singular, was first proved in [Kusuoka 1989]. Most of the arguments in Section 5.3 are from [Ben Bassat et al. 1999], where Kusuoka's result is extended to a larger class of Laplacians. The simple proof of Theorem 5.3.2 is new, but it doesn't prove the full singularity result. The computation of the Kusuoka measure on the unit disk was mentioned in [Strichartz 1999b]. Some results about the Laplacian Δ_ν defined with respect to the Kusuoka measure may be found in [Teplyaev 2004], [Teplyaev *].

Fractal blow-ups are defined in [Strichartz 1998]. Fractafolds are defined in [Strichartz 2003a]. Liouville-type theorems on blow-ups of SG are discussed in [Strichartz 1999a]. An explicit description of the spectrum of a compact fractafold based on SG is given in [Strichartz 2003a], including an explicit example of an isospectral pair constructed with the assistance of Robert Brooks. The description of the spectrum of the Laplacian on blow-ups of SG is from [Teplyaev 1998]. The results on singularities in Section 5.5 are from [Ben-Gal et al. *]. The results in Section 5.6 on quasi-elliptic operators is from [Bockelman & Strichartz *]. The rest of the material on products of fractals is from [Strichartz 2005a], which contains many other results of a more technical nature.

The results in Section 5.7 on solvability of differential equations are from [Strichartz 2005c]. See [Falconer 1999] and [Falconer & Hu 1999] for solvability of some nonlinear differential equations. The solvability of constant coefficient elliptic linear PDEs on arbitrary domains in Euclidean space is from [Malgrange 1955–56] and [Ehrenpreis 1961]. Heat kernel estimates for SG were first proved in [Barlow & Perkins 1988]. They were extended to nested fractals in [Kumagai 1993], to affine nested fractals in [Fitzsimmons et al. 1994], and to homogeneous heirarchical fractals in [Barlow & Hambly 1997]. An exposition of the probabilistic methods used may be found in [Barlow 1998]. See also [Grigor'yan & Telcs 2001], [Hambly & Kumagai 1999], and [Sabot 2003]. A proof of just the on-diagonal estimates using analytic techniques is presented in [Kigami 2001]. Similar heat kernel estimates on Sierpinski carpets, which are not pcf, are proved in [Barlow & Bass 1989, 1992, 1999]. See [Barlow et al. *] for a general stability result about heat kernel estimates. An interesting application of the heat kernel estimates on SG to Riemannian geometry is given in [Barlow et al. 2001]. The more refined description of the trace of the heat kernel (5.8.12) is from [Allan & Strichartz *], which also has numerical evidence for the periodic factor for the heat kernel on the diagonal. Similar numerical evidence concerning normal derivatives of the heat kernel was presented in [Ben-Gal et al. *]. Applications of heat kernel estimates in the product setting are given in [Strichartz 2005a].

The results on convergence of Fourier series on SG in Section 5.9 are from [Strichartz 2005b]. The extension to homogeneous hierarchical fractals is given in [Drenning & Strichartz *]. The generic transplantation theorem is from [Duong et al. 2002]. The asymptotics of the eigenvalue counting function is from [Kigami & Lapidus 1993].

The convergence of $s_{N_m} f$ to f for $N_m = \#V_m$ was first suggested by numerical evidence in [Oberlin et al. 2003] that the associated Dirichlet kernels do look like approximate identities. In fact, the values of $D_{N_m}(x, y)$ were extremely small for all pairs x, y with $R(x, y) \geq \delta_m$, with δ_m decreasing with m for very moderate values of m. It was conjectured there that these Dirichlet kernels satisfy estimates akin to the heat kernel. Note that the proof of Theorem 5.9.1 does not imply any such information.

The convergence of Fourier series on SG seems to be the first example of a result in fractal analysis that is actually stronger than the analogous result in smooth analysis. A second example, both closely related yet entirely different, is reported in [Strichartz *]. There it is shown that mock Fourier series, discovered in [Jorgensen & Pedersen 1998] and further studied in [Strichartz 2000b] and [Łaba & Wang 2002], also enjoy the same convergence property. These are expansions on certain Cantor sets in certain orthonormal bases of ordinary exponentials $\{e^{2\pi i \lambda x}\}$ for $\lambda \in \Lambda$, a certain discrete "spectrum." In fact, these exponentials are not eigenfunctions of any known Laplacian. The spectrum Λ also has large gaps, and once again if you take partial sums up to a gap, you get uniform convergence on the Cantor set for any continuous function. In this case the proof is based on showing directly that the associated Dirichlet kernels do behave like approximate identities on the Cantor set.

References

[Adams et al. 2003]   B. Adams, S. A. Smith, R. Strichartz and A. Teplyaev, *The spectrum of the Laplacian on the pentagasket*, Trends in Mathematics: Fractals in Graz 2001, Birkhauser, Boston, 1–24.

[Allan & Strichartz *]   A. Allan and R. Strichartz, *Spectral operators on the Sierpinski gasket*, in preparation.

[Bandt & Retta 1992]   C. Bandt and T. Retta, *Topological spaces admitting a unique fractal structure*, Fund. Math. **141**, 257–268.

[Barlow 1998]   M. Barlow, *Diffusion on Fractals*, Lecture Notes Math., Vol. 1690. Springer.

[Barlow & Bass 1989]   M. T. Barlow and R. F. Bass, *Construction of Brownian motion on the Siepinski carpet*, Ann. Inst. H. Poincaré Probab. Statist. **25**, 225–257.

[Barlow & Bass 1992]   M. T. Barlow and R. F. Bass, *Transition densities for Brownian motion on the Sierpinski carpet*, Probab. Theory Related Fields **91**, 307–330.

[Barlow & Bass 1999]   M. T. Barlow and R. F. Bass, *Brownian motion and harmonic analysis on Sierpinski carpets*, Canad. J. Math. **51**, 673–744.

[Barlow et al. *]   M. Barlow, R. Bass and T. Kumagai, *Stability of parabolic Harnack inequalities on metric measure spaces*, J. Math. Soc. Japan, to appear.

[Barlow et al. 2001]   M. Barlow, T. Coulhon and A. Grigor'yan, *Manifolds and graphs with slow heat kernel decay*, Invent. Math. **144**, 609–649.

[Barlow & Hambly]   M. T. Barlow and B. M. Hambly, *Transition density estimates for Brownian motion on scale irregular Sierpinski gaskets*, Ann. Inst. H. Poincaré **33**, 531–557.

[Barlow & Kigami 1997]   M. Barlow and J. Kigami, *Localized eigenfunctions of the Laplacian on p.c.f. self-similar sets*, J. London Math. Soc. **56**, 320–332.

[Barlow & Perkins 1988]   M. Barlow and E. Perkins, *Brownian motion on the Sierpinski gasket*, Probab. Theory Related Fields **79**, 543–623.

[Ben-Bassat et al. 1999]   O. Ben-Bassat, R. S. Strichartz and A. Teplyaev, *What is not in the domain of the Laplacian on a Sierpinski gasket type fractal*, J. Funct. Anal. **166**, 197–217.

[Ben-Gal et al. *]   N. Ben-Gal, A. Shaw-Krauss, R. Strichartz and C. Young, *Calculus on the Sierpinski gasket II: point singularities, eigenfunctions, and normal derivatives of the heat kernel*, Trans. Amer. Math. Soc., to appear.

[Bockelman & Strichartz *]   B. Bockelman and R. Strichartz, *Partial differential equations on products of Sierpinski gaskets*, preprint.

[Bougerol 1985]   P. Bougerol, *Limit theorems for products of random matrices*, in Products of Random Matrices with Applications to Schrödinger Operators (P. Bougerol and J. Lacroix, ed.), Birkhauser, Basel, Boston, pp. 1–180.

[Cucuringu & Strichartz *]   M. Cucuringu and R. Strichartz, *Self-similar energy forms on the Sierpinski gasket with twists*, in preparation.

[Dalrymple et al. 1999]   K. Dalrymple, R. Strichartz and J. Vinson, *Fractal differential equations on the Sierpinski gasket*, J. Fourier Anal. Appl. **5**, 203–284.

[Doyle & Snell 1984]   P. G. Doyle and J. L. Snell, *Random Walks and Electrical Networks*, Math. Assoc. Amer., Washington, D.C.

[Drenning & Strichartz *]   S. Drenning and R. Strichartz, *Spectral decimation on Hambly's homogeneous, hierarchical gaskets*, in preparation.

[Duong et al. 2002]   X. T. Duong, E. M. Ouhabaz and A. Sikora, *Plancherel type estimates and sharp spectral multipliers*, J. Funct. Anal. **196**, 443–485.

[Ehrenpreis 1961]   L. Ehrenpreis, *A fundamental principle for systems of linear differential equations with constant coefficients, and some of its applications*, Proc. Intern. Symp. on Linear Spaces, Jerusalem, 161–174. Jerusalem Academic Press, Jerusalem; Pergamon, Oxford.

[Falconer 1999]   K. J. Falconer, *Semilinear PDEs on self-similar fractals*, Comm. Math. Phys. **206**, 235–245.

[Falconer & Hu 1999]   K. J. Falconer and J. Hu, *Nonlinear elliptic equations on the Sierpinski gasket*, J. Math. Anal. Appl. **240**, 552–573.

[Fitzsimmons et al. 1994]   P. J. Fitzsimmons, B. M. Hambly and T. Kumagai, *Transition density estimates for Brownian motion on affine nested fractals*, Comm. Math. Phys. **165**, 595–620.

[Fukushima et al. 1994]   M. Fukushima, Y. Oshima and M. Takeda, *Dirichlet Forms and Symmetric Markov Processes*, de Grutyer Studies in Math. Vol. 19, de Gruyter, Berlin.

[Fukushima & Shima 1992]   M. Fukushima and T. Shina, *On a spectral analysis for the Sierpinski gasket*, Potential Anal. **1**, 1–35.

[Gibbons et al. 2001]   M. Gibbons, A. Raj and R. Strichartz, *The finite element method on the Sierpinski gasket*, Constructive Approx. **17**, 561–588.

[Goldstein 1987]   S. Goldstein, *Random walks and diffusions on fractals, Percolation theory and ergodic theory of infinite particle systems*, (H. Kesten, ed.), IMA Math. Appl., vol. 8, Springer, pp. 121–129.

[Grigor'yan & Telcs 2001]   A. Grigor'yan and A. Telcs, *Sub-Gaussian estimates of heat kernels on infinite graphs*, Duke Math. J. **109**.

[Hambly 1992]   B. M. Hambly, *Brownian motion on a homogeneous random fractal*, Probab. Theory Related Fields **94**, 1–38.

[Hambly 1997]   B. M. Hambly, *Brownian motion on a random recursive Sierpinski gasket*, Ann. Probab. **25**, 1059–1102.

[Hambly 2000]   B. M. Hambly, *Heat kernels and spectral asymototics for some random Sierpinski gaskets*, Fractal Geometry and Stochastics II (C. Bandt et al., eds.), Progress in Probability, Vol. 46, Birkhäuser, Boston, pp. 239–267.

[Hambly & Kumagai 1999]   B. M. Hambly and T. Kumagai, *Transition density estimates for diffusion processes on post critically finite self-similar fractals*, Proc. London Math. Soc. (3) **78**, 431–458.

[Hambly et al. *]   B. M. Hambly, V. Metz and A. Teplyaev, *Admissible refinements on finitely ramified fractals*, Proc. London Math. Soc., to appear.

[Hattori et al. 1987]   K. Hattori, T. Hattori and T. Watanabe, *Gaussian field theories on general networks and the spectral dimensions*, Progr. Theoret. Phys. Suppl. **92**, 108–143.

[Hutchinson 1981]   J. E. Hutchinson, *Fractals and self similarity*, Indiana Univ. Math. J. **30**, 713–747.

[Jonsson 1996]   A. Jonsson, *Brownian motion on fractals and function spaces*, Math. Z. **222**, 495–504.

[Jonsson & Wallin 1984]   A. Jonsson and H. Wallin, *Function spaces on subsets of* **R**^{n}, Math. Reports **2**, 1–221.

[Jorgensen & Pedersen 1998] P. E. T. Jorgensen and S. Pedersen, *Dense analytic subspaces in fractal L^2 spaces*, J. Anal. Math. **75**, 185–228.

[Kigami 1989]   J. Kigami, *A harmonic calculus on the Sierpinski spaces*, Japan J. Appl. Math. **6**, 259–290.

[Kigami 1993]   J. Kigami, *Harmonic calculus on p.c.f. self-similar sets*, Trans. Amer. Math. Soc. **335**, 721–755.

| [Kigami 1994a] | J. Kigami, *Effective resistances for harmonic structures on p.c.f. semi-similar sets*, Math. Proc. Cambridge Phil. Soc. **115**, 291–303. |

[Kigami 1994b]   J. Kigami, *Laplacians on self-simlar sets (analysis on fractals)*, Amer. Math. Soc. Transl. Ser. 2 **161**, 75–93.

[Kigami 1998]   J. Kigami, *Distributions of localized eigenvalues of Laplacians on p.c.f. self-similar sets*, J. Funct. Anal. **156**, 170–198.

[Kigami 2001]   J. Kigami, *Analysis on Fractals*, Cambridge University Press, Cambridge, UK.

[Kigami 2003]   J. Kigami, *Harmonic analysis for resistance forms*, J. Funct. Anal. **204**, 399–444.

[Kigami & Lapidus 1993] J. Kigami and M. L. Lapidus, *Weyl's problem for the spectral distribution of Laplacians on p.c.f. self-similar fractals*, Comm. Math. Phys. **158**, 93–135.

[Kigami et al. 2000]   J. Kigami, D. Sheldon and R. Strichartz, *Green's functions on fractals*, Fractals **8**, 385–402.

[Kumagai 1993]   T. Kumagai, *Estimates of the transition densities for Brownian motion on nested fractals*, Probab. Theory Related Fields **96**, 205–224.

[Kusuoka 1987]   S. Kusuoka, *A diffusion process on a fractal, Probabilistic Methods on Mathematical Physics*, Proc. of Taniguchi International Symp. (Katata & Kyoto, 1985) (Tokyo) (K. Ito and N. Ikeda, eds.), Kinokuniya, pp. 251–274.

[Kusuoka 1989]   S. Kusuoka, *Dirichlet forms on fractals and products of random matrices*, Publ. Res. Inst. Math. Sci. **25**, 659–680.

[Kusuoka & Zhou 1992]   S. Kusuoka and X. Y. Zhou, *Dirichlet forms on fractals: Poincaré constant and resistance*, Probab. Theory Related Fields **93**, 169–196.

[Laba & Wang 2002]   I. Laba and Y. Wang, *On spectral Cantor measures*, J. Functional Anal. **193**, 409–420.

[Lindstrøm 1990]   T. Lindstrøm, *Brownian motion on nested fractals*, Mem. Amer. Math. Soc. **420**.

[Malgrange 1955–56]   Malgrange, B., *Existence et approximation des solutions des solutions des équations aux dérivées partielles et des équations de convolution*, Ann. Inst. Fourier Grenoble **6**, 271–355.

[Metz 1993]   V. Metz, *How many diffusions exist on the Vicsek snowflake*, Acta Appl. Math. **32**, 227–241.

[Metz 1995]   V. Metz, *Hilbert projective metric on cones of Dirichlet forms*, J. Funct. Anal. **127**, 438–455.

[Metz 1996]   V. Metz, *Renormalization contracts on nested fractals*, J. Reine Angew. Math. **459**, 161–175.

[Metz 1997]   V. Metz, *Shorted operators: an appliation in potential theory*, Linear Algebra Appl. **264**, 439–455.

[Metz 2003a]   V. Metz, *The cone of diffusions on finitely ramified fractals*, Nonlinear Anal. **55**, 723–738.

[Metz 2003b]   V. Metz, *The Laplacian of the Hany fractal*, Arab. J. Sci. Eng. Sect. C Theme Issues 28 (IC) 199–211.

[Metz 2004]   V. Metz, *"Laplacians" on finitely ramified, graph directed fractals*, Math. Ann. **330**, 809–828.

[Metz 2005]   V. Metz, *The short-cut test*, J. Functional Anal. **220**, (2005), 118–156.

[Meyers et al. 2004]   R. Meyers, R. Strichartz and A. Teplyaev, *Dirichlet forms on the Sierpinski gasket*, Pacific J. Math. **217** 149–174.

[Mumford et al. 2002] D. Mumford, C. Series and D. Wright, *Indra's Pearls - The Vision of Felix Klein*, Cambridge University Press, Cambridge, UK.

[Needleman et al. 2004] J. Needleman, R. Strichartz, A. Teplayev and P.-L. Yung, *Calculus on the Sierpinski gasket I: polynomials, exponents and power series*, J. Funct. Anal. **215**, 290–340.

[Öberg et al. 2002]   A. Öberg, R. Strichartz and A. Yingst, *Level sets of harmonic functions on the Sierpinski gasket*, Ark. Mat. **40** 335–362.

[Oberlin et al. 2003] R. Oberlin, B. Street and R. Strichartz, *Sampling on the Sierpinski gasket*, Experiment. Math. **12**, 403–418.

[Peirone 2000]   R. Peirone, *Convergence and uniqueness problems for Dirichlet forms on fractals*, Boll. Unione Mat. Ital. Sez. B Artic. Ric. Mat. 3–B (8), 431–460.

[Peirone *]   R. Peirone, *Existence of eigenforms on fractals with three vertices*, preprint.

[Pelander & Teplyaev *] A. Pelander and A. Teplyaev, *Infinite dimensional i.f.s. and smooth functions on the Sierpinski gasket*, preprint.

[Rammal 1984]   R. Rammal, *Spectrum of harmonic excitations on fractals*, J. Physique **45**, 191–206.

[Rammal & Toulouse 1983] R. Rammal and G. Toulouse, *Random walks on fractal structure and percolation cluster*, J. Physique Letters **44**, L13–L22.

[Sabot 1997]   C. Sabot, *Existence and uniqueness of diffusions on finitely ramified self-similar fractals*, Ann. Sci. École Norm. Sup. (4) **30**, 605–673.

[Sabot 2003]   C. Sabot, *Spectral properties of self-similar lattices and iteration of rational maps*, Mem. Soc. Math. Fr. **92**.

[Shima 1991]   T. Shima, *On eigenvalue problems for the random walks on the Sierpinski pre-gaskets*, Japan J. Indust. Appl. Math. **8**, 127–141.

[Shima 1996]   T. Shima, *On eigenvalue problems for Laplacians on p.c.f. self-similar sets*, Japan J. Indust. Appl. Math. **13**, 1–23.

[Stanley et al. 2003]   J. Stanley, R. Strichartz and A. Teplyaev, *Energy partition on fractals*, Indiana Univ. Math. J. **52**, 133–156.

[Strichartz 1998]   R. Strichartz, *Fractals in the large*, Canad. J. Math. **50**, 638–657.

[Strichartz 1999a]   R. Strichartz, *Some properties of Laplacians on fractals*, J. Funct. Anal. **164**, 181–208.

[Strichartz 1999b]   R. Strichartz, *Analysis on fractals*, Not. Amer. Math. Soc. **46**, 1199–1208.

[Strichartz 2000a]   R. Strichartz, *Taylor approximations on Sierpinski gasket-type fractals*, J. Funct. Anal. **174**, 76–127.

[Strichartz 2000b]   R. Strichartz, *Mock Fouriers series and transforms associated with certain Cantor measures*, J. Anal. Math. **81** 209–238.

[Strichartz 2003a]   R. Strichartz, *Fractafolds based on the Sierpinski gasket and their spectra*, Trans. Amer. Math. Soc. **355**, 4019–4043.

[Strichartz 2003b]   R. Strichartz, *Function spaces on fractals*, J. Funct. Anal. **198**, 43–83.

[Strichartz 2005a]   R. Strichartz, *Analysis on products of fractals*, Trans. Amer. Math. Soc. **357**, 571–615.

[Strichartz 2005b]   R. Strichartz, *Laplacians on fractals with spectral gaps have nicer Fourier series*, Math. Res. Lett. **12**, 269–274.

[Strichartz 2005c]   R. Strichartz, *Solvability for differential equations on fractals*, J. Anal. Math. **96**, 247–267.

[Strichartz *]   R. Strichartz, *Convergence of mock Fourier series*, J. Anal. Math. (to appear).

[Strichartz & Usher 2000]   R. S. Strichartz and M. Usher, *Splines on fractals*, Math. Proc. Cambridge Philos. Soc. **129**, 331–360.

[Teplyaev 1989]   A. Teplyaev, *Spectral analysis on infinite Sierpinski gaskets*, J. Funct. Anal. **159**, 537–567.

[Teplyaev 2000]   A. Teplyaev, *Gradients on fractals*, J. Funct. Anal. **174**, 128–154.

[Teplyaev 2004]   A. Teplyaev, *Energy and Laplacian on the Sierpinski gasket*, Fractal Geometry and Applications: A Jubilee of Benoit Mandelbrot, Part 1. Proc. Sympos. Pure Math. **72**, 131–154.

[Teplyaev *]   A. Teplyaev, *Harmonic coordinates on fractals with finitely ramified cell structure*, Canad. J. Math., to appear.

Index